One of *Inverse*'s Best Sc

One of *Forbes*'s Twelve Best

More praise for

HOW TO WALK ON WATER AND CLIMB UP WALLS

"Hu masterfully explains scientific revelations such as how insects walk on water, snakes slither and mosquitos survive rainstorms using rich tales of discovery and analogies that brim with satisfying 'aha' moments."

—NATALIE PARLETTA, *Cosmos*

"Roboticists tasked with designing the machines of tomorrow are inspired by the spectacular blueprints created by nature. . . . David Hu dives into these all-natural plans and explains why they're so useful to robot design."

—SARAH SLOAT, *Inverse*

"A fascinating book."

—DOMINIC LENTON, *Engineering and Technology*

"With infectious enthusiasm and curiosity, David Hu asks why natural selection may have favored one design over another. From flying snakes to the eyelashes of giraffes, he sees mechanical challenges everywhere, and his crazy experiments help us understand how animals dry their bodies, move, pee, and eat."

—FRANS DE WAAL, author of *Are We Smart Enough to Know How Smart Animals Are?*

"I recommend this book to all interested in biology and nature-inspired engineering. They will learn much and be marvelously entertained."

—BERT HÖLLDOBLER, coauthor of *The Ants*

"Read this fascinating book to learn how sharks move, why an elephant doesn't take longer to empty its bladder than a human does, how mosquitoes fly in the rain, and how cockroaches avoid bumping into walls in the dark. Read it too to learn why the study of these seemingly obscure corners of the living world has the potential to offer enormous benefits to humanity."

—ROB DUNN, author of *Never Out of Season*

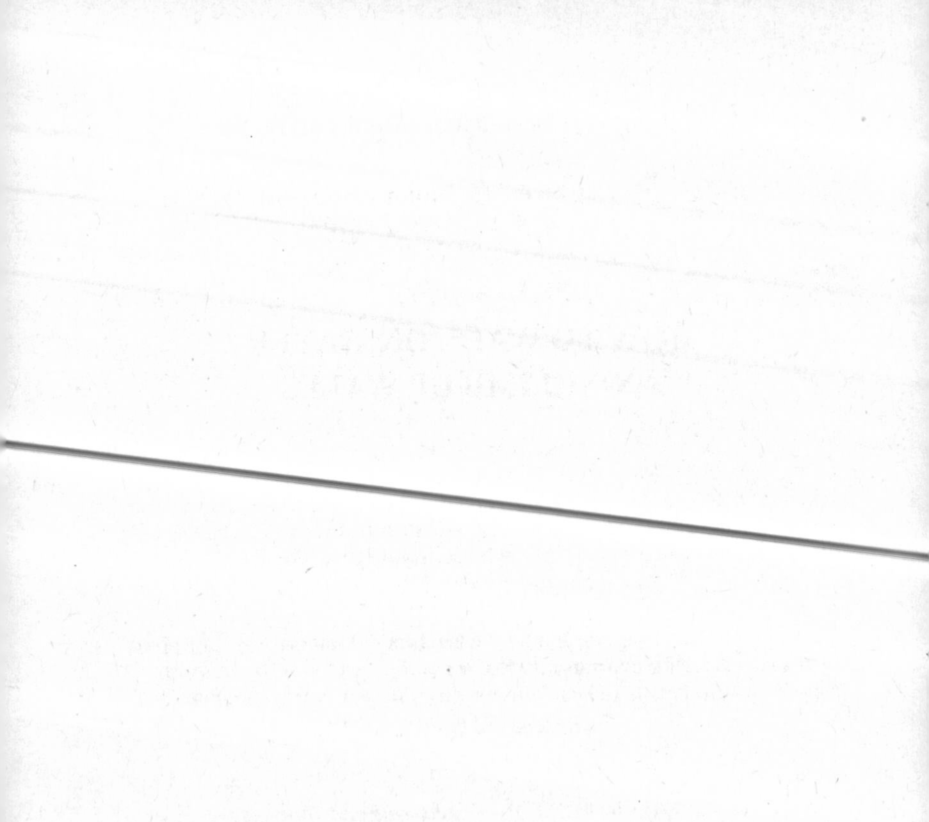

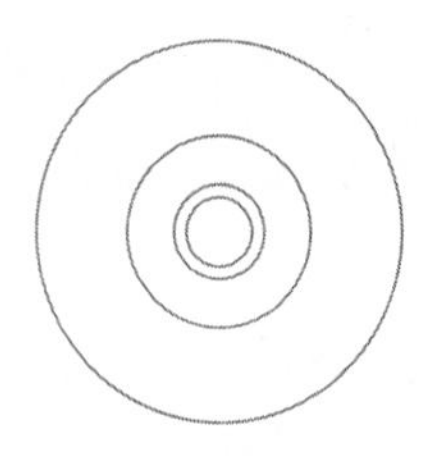

How to Walk on Water and Climb up Walls

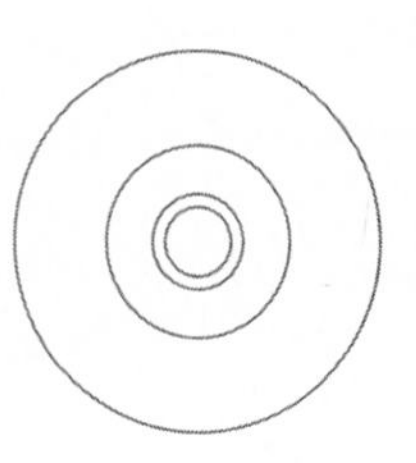

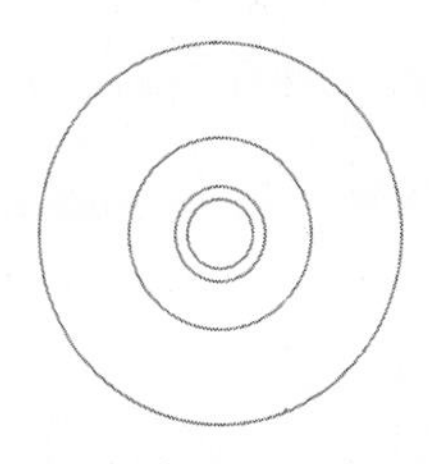

How to Walk on Water and Climb up Walls

Animal Movement and the Robots of the Future

DAVID L. HU

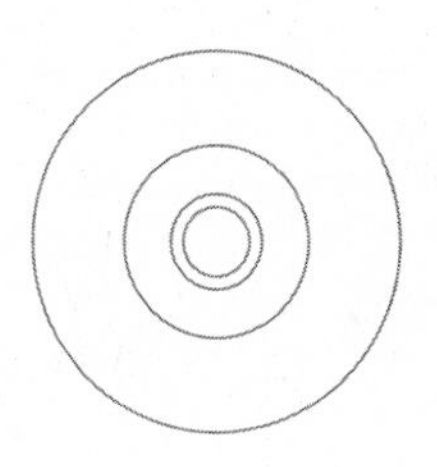

PRINCETON UNIVERSITY PRESS
Princeton and Oxford

Published by Princeton University Press
41 William Street, Princeton, New Jersey 08540
6 Oxford Street, Woodstock, Oxfordshire OX20 1TR

press.princeton.edu

Library of Congress Control Number: 2018949420
Second paperback printing, 2020
Paperback ISBN 978-0-691-20416-1
Cloth ISBN 978-0-691-16986-6

British Library Cataloging-in-Publication Data is available

Editorial: Alison Kalett and Lauren Bucca
Production Editorial: Kathleen Cioffi
Text Design: C. Alvarez
Jacket/Cover Design: Amanda Weiss
Production: Erin Suydam
Publicity: Sara Henning-Stout and Katie Lewis
Copyeditor: Jodi Beder

This book has been composed in Charis SIL

Printed in the United States of America

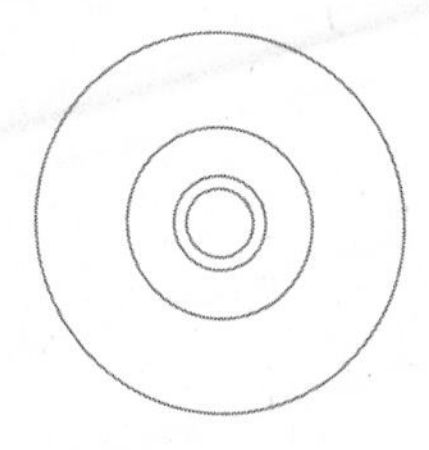

To Jia, for believing in me

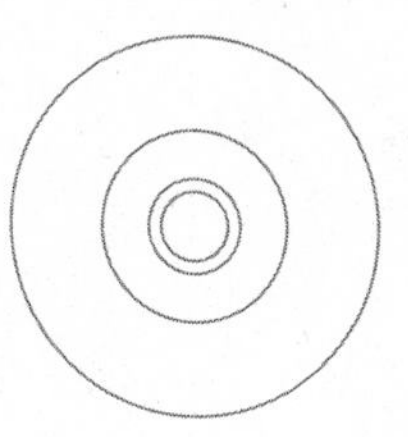

Contents

Acknowledgments

Many thanks to my editor Alison Kalett, who called me on February 2, 2015, to recruit me to write this book, and advised me to its completion. Thanks also to the knowledgeable staff at Princeton University Press, including Stephanie Rojas for marketing, Lauren Bucca for editing, Kathleen Cioffi for production editing, Dimitri Karetnikov for illustrations, Jodi Beder for copyediting, Virginia Ling for the index, and Sara Henning-Stout and Katie Lewis for publicity.

Thanks to the National Science Foundation, the Army Research Office, the Smithgall Watts Foundation, and Georgia Institute of Technology for funding my research, some of which is covered in this book. Thanks also to the thirty scientists who agreed to be interviewed and featured in the book, and a number of others who provided suggestions and comments, especially Jake Socha, who read and commented on the entire manuscript.

Thanks to Youmei Zhou for preparing the references and figures, and photographers Tim Nowack, Candler Hobbs, and Brian Chan for taking many of the color photos. Thanks to the Atlanta Zoo for caring for the pandas, elephants, and other animals, and helping me observe them safely over the last decade.

Lastly, a special thanks to two anonymous reviewers who together provided thirty pages of comments that improved the accuracy and readability of this book.

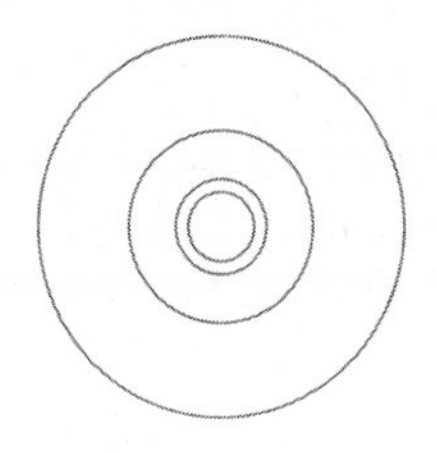

How to Walk on Water and Climb up Walls

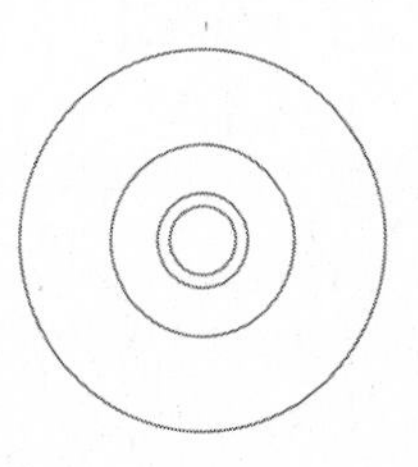

INTRODUCTION

The World of Animal Motion

When I met my wife for the first time, she brought along a brown toy poodle named Jerry. He was a Valentine's gift from her ex-boyfriend, and he proved the perfect subject for my next scientific experiment. I spent a great deal of time putting sticky notes on Jerry's fur and filming him with a high-speed camera. Jerry did not like his stickers very much and tried to bite them off. If the stickers were on the top of his head or neck, he had another method to remove them. He shook his body and head back and forth several times, making me take a step back (Plate 1). His brown curls flew, sending dust and fleas in every direction, along with my stickers. This neat little trick, called the wet-dog shake, looked like a silly, useless act.

As I analyzed the high-speed films, I learned that Jerry's shake generated accelerations up to 12 times earth's gravity, higher than the acceleration of a Formula One race car as it turns a corner. When I gave Jerry a bath, I found his shake could remove up to 70 percent of the water contained in his fur. It took only a fraction of a second, whereas our laundry machines take minutes to perform comparably. How is the wet-dog shake so effective?

My student Andrew Dickerson and I built a wet-dog simulator, a rotating column that spun a clipping of Jerry's fur at the same rates we observed. A camera was attached to the spinning frame so we could see drops releasing from patches of fur as it was spun. It was like we had a front-row seat to the water ejection process. On our wet-dog simulator, 12 times earth's gravity was the minimal acceleration to remove the smallest drops from the fur. This regime exactly coincided with the acceleration that Jerry was generating.

To find out if Jerry was alone in his ability to shake water, I scoured Atlanta for the biggest range of animals I could find, taking trips to laboratories on campus, the local park, and the Atlanta Zoo. Over the course of the next few years, the zoo would grow accustomed to strange research requests, such as, "Can we come film your pandas shaking off water?" Ultimately, we gathered high-speed films of animals across a 10,000-fold range in body mass, from a mouse to a bear. Bears shake at four times per second, dogs at four to seven times per second, rats at 18 times per second, and mice at a dizzying 29 times per second. The blink of a human eye would miss over 10 shakes of a mouse. Why do smaller animals shake more times per second? A smaller animals has a smaller radius, which causes it to generate less centripetal force. In order to generate the same drop-releasing forces as a larger animal, it must spin faster.

Jerry's owner brought me a lifetime of love and two wonderful children, who later would also become unknowing subjects in my experiments. I'll tell you more about our adventures in the coming chapters. But I give credit to Jerry the brown toy poodle for buying me a ticket to my scientific obsession, the world of animal motion.

* * *

Animals may look quite different, but they share one thing in common: they must move in order to survive. Movement evolved for a simple reason: the requirement for energy. It is one of the things that distinguishes animals from plants, which are generally sedentary. Plants are known as *autotrophs*, or self-nourishing, because they make their own food using sunlight. For them, large-scale movement is unnecessary, except for reproduction. In comparison, animals are *heterotrophs,* and obtain their energy by continually seeking out and consuming food. Plant-eaters move to forage, and predators move to give chase. Be it predator or prey, fast and responsive movements are one way to avoid being eaten. Yet, the more energy they expend moving, the more animals need to eat, to replenish their energy supply. Thus, animals often push the envelope in terms of speed, economy, and maneuverability.

Animal movement also involves navigating a variety of environments. It's easy to forget how difficult it is to deal with natural environments. We fly over them with ease in our airplanes, or drive miles and miles without a second thought. In contrast, consider the pigeon. A pigeon flying from A to B might encounter vortices or other turbulence in the air that can blow it off its intended course. The air it flies through is filled with obstacles such as the branches of a tree, which themselves might be rattled back and forth by the wind. You might think that such difficulties are specific to air travel, but animals on the ground have it just as hard. A salamander traveling from A to B might encounter twigs, forest litter, or mud on its way. It may even have to transition amphibiously from land to water in order to lay its eggs in a wet environment.

A changing world creates changing conditions for animals as they move. Animal movement must change from night to day, and from season to season. Animals face rain, sleet, and snow. Starting in spring, bees gather pollen, and they can become completely covered in it, as they go from flower to flower. In addition to being covered by inanimate materials, animals must deal with crowding by fellow members of their species. Consider a dense flock of pigeons or a school of fish: like us, they face traffic jams on the search for food.

Although moving from A to B is important, another type of movement occurs on a smaller scale, and is equally important for survival. Animals transport matter into and out of themselves, a critical part of eating and waste production. We don't often think about such movements because we live in a built world, having designed tools like spoons and shovels to transport matter. In nature, animals manipulate matter with their own body parts. A dog laps water using its soft tongue and an elephant picks up fruit using its long flexible trunk. Animals live in a world of parasites, such as ticks and fleas, and grooming can be a matter of life and death. One way to combat these parasites is through animal movement, often with body parts. The cat tongue is like a hairbrush, but it is also much more. It is covered in an array of sharp spikes containing a U-shaped cavity at the tip that can spontaneously imbibe saliva and release it on contact with the individual hairs that it encounters.

Animal motion is all around us. It is the principal way animals get things done in the world. How did such a diversity of animal movements come about?

The variety of animal motion is made possible by a single common thread, evolution. Evolution functions as a deceptively simple but powerful algorithm. Organisms reproduce, making imperfect copies of themselves—that is, offspring that are different from their parents. If some of this variation improves an organism's ability to survive or reproduce, and if this variation can be passed on to that organism's offspring, then over time the population will evolve and adapt. It's a simple idea, yet few imagine the sheer variety evolution has created over the 3.5 billion years it has been at work.

All types of animal movement, no matter how bizarre, emerged through the process of evolution. Walking on water is one fascinating example. Insects evolved around 400 million years ago, and 300 million years later, terrestrial insects and spiders began to colonize the water surface. This migration helped them to avoid predators, to find new sources of food and safe places to hatch their young. The most primitive surviving water-walkers, the *Velia* or water treaders, look quite similar to the bugs from which they evolved. Like many land insects, they use their six legs to walk in a manner similar to an ant, on the tips of their feet. This motion is not very effective on water, and they move frantically and with little progress, as if perpetually slipping on ice. Moreover, their slow gait restricts them to being close to shore. These primitive water-walkers inhabit shallows where duckweed and other emerging vegetation gives them areas to climb onto for safety.

Over time, the middle legs of the water-walkers evolved to be longer, conferring on the water-walkers a distinct advantage over walking on the tips of their feet. Eventually their legs became so long that they could act like oars. This new species, called the water strider, can row like a rowboat, balancing on its remaining legs as pontoons for support. It's a highly effective gait, and such insects are nearly impossible to catch by hand. In turn, the water striders became so specialized at moving on the water surface that they move slowly on land, awkwardly dragging their canoe-like legs behind them. There is no turning back for the water striders. The water surface has become their permanent home.

Invertebrates were not alone in evolving the ability to walk on water. The Jesus Christ lizard, or basilisk lizard, is green with white dots, and large yellow eyes. It moves like a normal lizard except that it can run on water in short bursts if frightened. It has long fringed toes that it uses to slap the water surface and support its weight. Similarly, the western grebe, a black and white bird with red eyes, also runs on water, despite weighing up to ten times as much as a basilisk lizard. In elaborate mating displays called "rushing," males run across water distances up to 50 meters in order to attract females. Even after mates are selected, male and female birds will run across water together to reinforce pair bonding. Both the lizards and the birds are constrained by their evolutionary origins. While the smallest vertebrates are insect-sized, the basilisk lizards and western grebes have internal bones, making them too large and heavy to walk on water effortlessly like water striders, which have exoskeletons.

Humans cannot overcome their evolutionary origins either: our feet are simply too small to support our weight upon the water surface. Leonardo da Vinci conceived of kayak-sized floats that could be worn on one's feet. With the aid of poles with floats at their tips, one could carefully cross-country ski on water. Even with the aid of such tools, however, we could never run across water like the basilisk lizard. Our muscles simply do not generate enough power to push the water fast enough to support our weight. Evolution is both a blessing and a curse: it causes many animals to specialize their locomotion in particular media, such as air, land, or water. Evolution, for example, has made us extremely energy-conserving when walking on land, but quite awkward when walking on water. Water striders have the opposite problem—elegant on water, and clumsy on land.

* * *

The study of animal motion is not new—in fact it has at least a four-hundred-year history. Questions about animal propulsion were around long before we had the equipment to carefully study them. One of the first to ask these questions was Leonardo da Vinci, who dissected animals and humans in secret. Leonardo's sketchbooks are filled with equal parts drills and helicopters and anatomical drawings of animals, as if they too were machines. At the time, many people believed in vitalism, that living things had a soul, which had never been observed or measured, yet made them inherently alive. In contrast, Leonardo took a logical rather than mysterious view of the world. His big idea was that the scientific method could make sense of anything around him, no matter how mysterious.

A book that is credited with drawing mathematicians and nonbiologists to animal motion was *On Growth and Form*, written in 1915, but delayed in publication till 1917 due to the First World War. Its author was D'Arcy Thompson, a Scottish biologist and pioneer of mathematical biology. His thesis was that biologists had not sufficiently emphasized the influence of mechanics and physical laws on the shapes and growth of organisms. His book provided mathematical descriptions of the shape of fish, birds, and mammals. The book was flawed because Thompson had not accepted Darwin's theory of natural selection. Nevertheless, it inspired generations of artists and scientists to cross disciplines, to begin using mathematics to describe the shapes of animals. A new take on animal morphology began, where biologists used mathematics to describe the shape of animals such as fish, which evolved into shapes as different as an elongated pike and a flat flounder. Describing these shapes was an important first step to understanding animal motion because the shapes of animals greatly influence the forces they feel and create as they move through fluids.

Biologists continued to use mechanics, giving rise to a new discipline called *biomechanics*. Cambridge zoologist Sir James Gray is arguably the father of modern biomechanics, the study of the physics of animal movement and shape. Today, biomechanics is not as clear-cut an area as it was then, as it now intersects with studies of microscopic things such as the biophysics of the cell and with human-centered topics such as exercise physiology. But for the purpose of this book, the biomechanics of animals is a field that indeed began with Gray. In the 1930s, Gray conducted the first studies of the swimming of fish and dolphins. In calculating the power that a dolphin would need to swim, he came to the startling conclusion that dolphins should not be able to swim. This controversy, which became known as Gray's Paradox, continued to draw non-biologists to the study of animal motion. It drew mathematicians, such as Sir James Lighthill, who were interested in understanding the body motions that made fish swim quickly or with low energy use.

The study of animal motion was energized by the advent of technology that enabled sharper images of animals. In the 1930s, electrical engineer Harold Edgerton popularized the use of strobe photography to capture high-speed motion. Decades later the commercialization of computers led to the use of high-speed cameras and computer algorithms that could automatically track fish and the fluid flows behind them. The digital era led to further progress in robotics, and new manufacturing technologies like 3D printing. The latter is an especially important tool that is becoming widespread among scientists studying animal movement. We will discuss 3D printing in the context of artificial shark scales in Chapter 4.

Today, the gradual merging of scientific disciplines has been a key part of progress in animal motion research. Fluid mechanics students eagerly take courses in fish anatomy. Roboticists designing robotic climbers read the classical work of German physiologists who first studied how insects grip surfaces. Material scientists bring biomaterials like oyster shells into their labs to crush them and examine them with microscopes. In many ways, biological techniques and practices are becoming more widely accepted and of greater interest to other fields. These fields in turn are infusing biology with new ideas and high-tech tools, leading to ways of doing science that were not possible just twenty years ago. In my mind, this marks a turning point in how animal motion is studied.

The goal of this book is to introduce readers to the world of animal movement and to the scientists who study it. I pay particular attention to key concepts that scientists use to understand the diversity of animal movement, showing that just a few physical concepts can give an intuitive understanding of a number of animal shapes and movements. Along the way, I hope to show that studying animal movement will lead to solutions to difficult problems of societal importance, such as designing more effective propellers or building robots to care for the elderly.

Most books on animal movement characterize animals according to the media that they live in, such as air, water, or land. Others separate animals into gait types such as walking, jumping, swimming, or flying. I've avoided these approaches, and instead focus on the broad physical principles at work. By focusing on principles, I can group seemingly disparate animals together. This is the rationale behind grouping sharks and eyelashes together in Chapter 4: they both involve fine surface textures that affect flows. Focusing on physical principles allows me to group animals and robots together. This is the rationale for putting walking robots and swimming fish together in Chapter 5: they both transfer energy in ways that reduce their energy use in propulsion. Another way to say this is that they have high fuel economy or gas mileage. However, I will avoid using the word "efficiency" in this text because engineers considering efficiency to only be non-zero if an animal is climbing a hill. Only then is an animal doing work against a force, in this case gravity. Going on flat ground at constant speed can't be described by efficiency because it doesn't actually require doing any work. Although this idea may be against your intuition, I will discuss how Newton's laws make it the case in Chapter 5. I believe a focus on principles rather than appearance will lead to thinking about animals in new and useful ways.

As you will see, this book is heavily focused on fluid mechanics, the physics of fluid motion such as the motion of air and water. Water constitutes more than 70 percent of the surface of our planet. As a result, a good number of animals have adapted to move through it. At the same time, many animals are composed of 70 percent water and must consume water on a daily basis. As a result, animals possess a number of internal bodily processes to facilitate the motion of fluids on their insides. Animals produce fluids such as saliva to moisten their food, or urine to excrete wastes. We'll discuss urination in Chapter 3 when we talk certain body and organ shapes that are effective at driving fluid flow.

I have also structured this book such that each chapter focuses on the stories of several key scientists. My goal is to convey the scientific process as if it were a mystery story. When I was younger, I loved reading mystery books by Agatha Christie because I could follow her logic as she presented the details of each case. I also enjoyed reading about the strange characters that brought the case to life. As we follow the scientists, we'll encounter a number of characters. Sometimes these characters are the machines the scientists are working with, the wind tunnels and high-speed cameras. Other times, they are the team that is at work, with members from biology, engineering, and physics. I've done my best to showcase the team effort that is necessary to make science happen, but I often leave the spotlight on the main character for the sake of story. I've chosen studies that were done early in a scientist's career, so you can see what it's like to embark on new territory. Since I've featured studies published in the years 2000 to 2018, a good number of these scientists may still be actively working, depending on when you are reading this book. Of course, no book on animal movement would be complete without the animals themselves, which are sometimes not so cooperative in revealing their secrets.

American astronomer Carl Sagan once said, "Science is a way of thinking much more than it is a body of knowledge." I hope that by following the journeys of a few select scientists, I can convey how they approach problems in animal motion. The details of this journey and the moment of discovery are often not contained in the scientists' journal papers, which are cited in the bibliography. Instead, I came upon the material over a series of interviews I conducted with the scientists from 2015 to 2017. Moreover, all the material in this book has been read and verified by the featured scientists themselves. For them, the scientific process involves a number of key steps. They conceive of the idea that they want to test, and work to hone their idea into a well-defined question. Then they design and perform an experiment that answers this question. When possible, they follow up by building a device that tests the original concept. Headway is made by a combination of logical thinking, help from team members, hard work, and serendipity. How much of their success in making discoveries is indeed luck? The answer is, less than you would think. Louis Pasteur said, "In the fields of observation, chance favors only the prepared mind." I try to slow down the moment of discovery as much as possible, so that you can see the logical steps that led to this aha! moment. This moment of discovery is not so much a moment of genius, but a series of logical steps that led to a conclusion.

In this book, and in the field of animal movement, we take a departure from much modern biology, which is focused on cellular and molecular science and model organisms, such as yeast, fruit flies, and mice. These animals are used because much is known about them, making tightly controlled studies possible. In contrast, scientists in this book study animals for the opposite reason, because so little is known about them. The flying snake must be captured in the rain forests of Singapore, the humble cockroach must be raised in the lab and fed like a pet. Be it an obscure animal or a disgusting one, these animals can reveal physical principles that can lead to broader claims about movement, relevant not just to animals but also to robots.

The word *robota*—in Czech, literally, "forced laborer"—was introduced in 1920 by Czech playwright Karel Čapek. Since the computer era, the field of robotics has expanded rapidly, and has been tasked with working beyond the factory floor. Automation is particularly difficult on rugged terrain, which covers much of our planet. Even the insides of our houses are filled with a variety of terrain, including carpet, hardwood floors, piles of clothes, and children's toys. These places are too unpredictable, too ridden with obstacles for wheels to work. Instead, some believe a legged robot might be the best strategy for movement. And if we are looking to legs, studies of animals can be of great use.

Studying animals can also teach us about the importance of body size, which can in turn influence how we design machines. In physics, bigger is different. As animals grow, certain forces that were previously negligible become important. For example, large animals simply cannot afford to fall down because they will be easily hurt. But as animals get smaller, their bones appear to be effectively stronger due to the physics of scaling. This is why fleas can jump 120 times their body length without being injured. Such invincibility at small sizes gives small animals a wider range of behaviors. Small animals are naturally more robust, so much so that they regularly crash into objects without damage. These fundamental issues of animal movement apply to machines as well. By understanding how animals of different sizes move differently, we can gain intuition into how to design machines of different size.

* * *

I start this book with my own beginning in the study of animal motion, investigating how insects can walk on water, the subject of my doctoral thesis. For this book, this seemed like a good place to start because such insects are so common. Across ponds, lakes, and streams, they leisurely stand on the water surface. We don't give them a second thought, but clearly they are a feat of nature. To understand the water striders, I learned to use a high-speed camera, a tool important to the study of animal movement that captures movements too fast for the human eye to see. I'll introduce the concept of surface tension, a tendency of a fluid surface to minimize its surface area. Surface tension is responsible for forming the shape of a water drop and supporting an insect's weight as it walks on water. I'll also introduce the concept of surface texture, in examining the hairy coating on the water strider's legs that makes the water strider water-repellent. Along the way to understanding these insects, I met a talented mechanical engineer, who built a robotic water strider. Chapter 1 sets the stage for the rest of the book in that it begins with a simple question about animal movement, and finishes with a proof-of-concept, here the construction of a device that can walk on water. It also shows how welcoming the field of animal motion can be to those with no training in biology.

Chapter 2 begins after my doctoral thesis, when I went to New York City to spend my postdoctoral years studying the motion of snakes. There, I learned about how solid surfaces can interact with each other in surprising ways. It is this frictional interaction that permits a snake to slither effortlessly over carpet and other seemingly uniform surfaces. Other animals can slither through sand and soil, as if they are swimming through water. I placed this chapter after walking on water because moving through soil is just as incredible as walking on water. It takes us hours to dig a hole in soil just a few feet deep. However, animals with long, streamlined bodies can easily dive down through sand. In this chapter, we will also learn about another tool, X-ray high-speed video, that can image animals underground.

One of the important concepts in Chapter 2 is the idea of optimality, that certain body shapes are well suited for moving through a particular medium. This streamlined body shape makes motion through sand and mud possible. In Chapter 3, we dig deeper into optimality, discussing three animals with particular body shapes that are best for performing some desired function. Of course, the process of evolution is not goal-directed, and animals carry with them a number of constraints that make it impossible to achieve perfection, also known as a *global optimum*. However, in the examples presented, we find that animals are good at dealing with the cards that they have been dealt.

Now that we have zoomed out and looked at the broad form of animals, it's time to zoom back in, to the world of the small, in Chapter 4. We are not accustomed to looking at small features on large objects. For instance, a car's shape is recognizable, but how often do we hold a magnifying glass to its surface? This is where nature departs from the built world. An animal grows by the reproduction of individual cells, which generates not just the animal's overall shape, but also intricate patterns on its surface. Growth is how a shark develops the fine scales on its surface, and how you grow eyelashes to protect your eyes. I'll talk about the hydrodynamic properties of each of these fine structures.

One of the drivers of animal motion is the need to move with the highest fuel economy possible, to save energy for other activities. Escape strategies, on the other hand, have a different need: speed, as in the fast escape stroke of the water strider or the C-start, the snapping of a fish's body like a whip when it is startled. These body motions, like sprints, involve high accelerations that rapidly convert fluid motion into heat. Nevertheless, being economical is important for any animal that must travel long distances. In Chapter 5, I discuss animals that can move using very little energy. I will introduce the concept of *energy transfer*, which is the main way that animals reduce their energy consumption. We do the same when we walk: our legs act like pendulums that transfer energy from gravitational to kinetic energy. Fish push this idea to its limits by harvesting energy from their surroundings similarly to the way a kite is pushed by the wind to move.

Thus far, we had not considered animals' interactions with obstructions and other unfavorable conditions in their environment. Our built world tries to remove many of the obstructions around us to facilitate transportation; highways have been designed as smooth and straight as possible. In comparison, when a bee flies through a field to gather pollen, it is surrounded by thousands of plant stems, each of them waving in the wind. The bee's solution is hard to believe. It simply crashes into the stems over and over on its way to find pollen. Its wings have special crush-zones that store elastic energy like springs, bending without breaking. In Chapter 6, we'll also talk about other injury-preventing strategies of insects, such as how mosquitoes survive a rainstorm.

I have focused so far on adaptations we can see, but in Chapter 7 we turn to adaptations that are under the hood—the nervous system. The nervous system is put to the test in particular in insect flight, where one of the most difficult tasks is to stay motionless, or hover. It is difficult for a fruit fly to hover, because the fruit fly's body is inherently unstable. Like a sheet of paper, the fruit fly does not tend to fall straight when dropped. The fly is affected by the air currents that it itself generates as it falls. The nervous system works with the body to put hovering and other kinds of locomotion under automatic control. Automation makes animal motion repeatable and robust, with as little input from the animal as possible, like driving with cruise control.

Thus far, we have considered how individual animals move. This approach is sufficient for solitary animals, but a good number of animals live in groups, the subject of Chapter 8. Flocks of starlings, packs of wolves, and colonies of ants are all examples of groups of animals that cooperate. Cooperation is a key innovation in the evolution of animals. Cooperation is so advantageous that once it evolves in an animal, it is there to stay. We discuss cooperation of fire ants and the engineering of cooperation in a swarm of 1000 robots.

I close the book with my thoughts about the future of animal motion. We are standing at an exciting point where changes are occurring very rapidly, both within the study of animal movement and in adjoining fields. Animals can now be captured with 3D technology, capturing both their bodies and the positions of their bones. Technology is enabling robots to start to move in a lifelike manner, and at a size comparable to animals. Micro-fabrication is making possible insect-sized flying robots. Snakelike robots are being used in search-and-rescue operations. New kinds of robots called biohybrids are composed of actual rat muscle tissue, but grown into un-ratlike shapes like the manta ray. With these exciting advances going on, I'll discuss a few simple things you can do to participate in discovery and to help others better understand and appreciate the field of animal motion research.

I hope you are as excited as I am to take a bird's-eye view of animal movement. Writing about each of these areas in the book has changed me, as it has changed many of my colleagues. The German scientist Haike Vallery once told me that as soon as she started studying walking, she began to walk very slowly, considering every single move she made. I hope that as you read each chapter, it too will change the way you consider the world. Remember, science is not about answers. It is about a careful level of inquiry, a curiosity about the way the world works. Take this curiosity with you as you begin to explore the wide world of animal movement.

CHAPTER 1

Walking on Water

On calm ponds and rivers, long-legged insects stand on the water surface as if it is land (Plate 2). They are the water striders, and in this chapter, we'll learn how their motion has inspired a water-walking robot. From above, the bodies of water striders are dark brown and kayak-shaped, streamlined for speed. Although they stand quite still most of the time, at the first sign of an intruder, they seem to just disappear, generating the equivalent of a sonic boom on the water surface, an explosion of small water waves radiating outward behind them. How can they stand on water when we cannot?

Water striders can stand on water because their small size allows them to take advantage of surface tension forces. Conversely, at your body size, the surface of water appears quite weak. You can step into a bathtub or stick straws into a drink without a second thought. However, water striders are small and light enough that they are sensitive to forces that you would find negligible. Specifically, they exploit surface tension forces, which arise due to the attraction of water molecules to each other. It takes energy to push these molecules apart, or to increase the water's surface area. Because of surface tension, bubbles are round to minimize their surface area. At the same time, a water

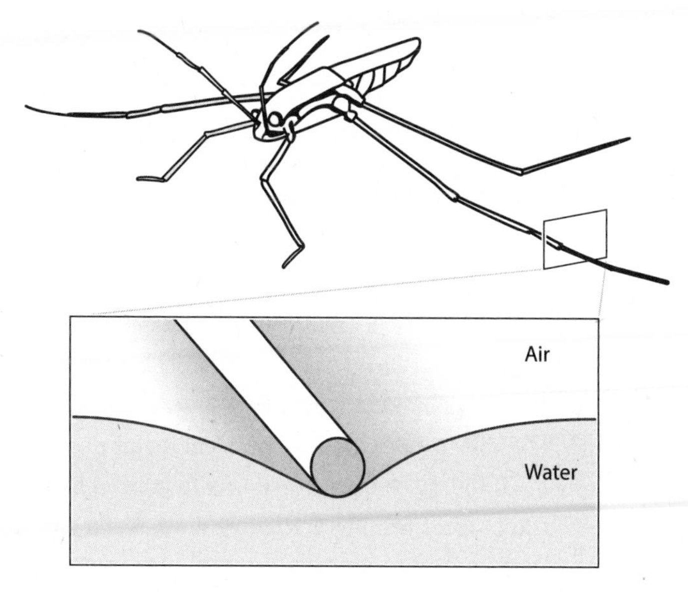

FIGURE 1.1. A schematic of a water strider standing on the water surface. The weight of the strider is supported by the surface tension of the water, which causes the water surface to deform like a trampoline. The inset shows the cylindrical leg depressing the water surface.

strider's low weight of only 10 milligrams, or just three sesame seeds, is enough to bend but not break water surfaces. Surface tension supports the water strider's weight just like a trampoline supports yours (Fig. 1.1). Water striders experience a different world than you do because of the physical consequences of *scaling*. Simply put, the forces acting on an animal depend upon its body size. We'll see this principle at play throughout this book, but it especially makes a difference for small animals.

If you were to try walk on water like a water strider, your body weight, about 10,00,000 times that of a water strider, would require feet nearly 10 kilometers across to be supported by surface tension. No one has ever built a foot that big, but there have been attempts to make shoes the size of kayaks, which are supported by the same buoyant force that prevents boats from sinking. When such buoyant forces act on feet, the result is often ungainly. In annual competitions where students wear boats on their feet and attempt to walk on water, forward motion is at best two steps forward, one step back.

I was first introduced to water striders by John Bush, a mathematician who at the time was one of the world's experts in fluid flows driven by surface tension. In the fall of 2001, I enrolled in his introductory course in fluid mechanics, which culminated in a final course project of the student's own design. When I came to his office one day, he took a book off his shelf and pointed to a page that stated that infant water striders shouldn't be able to move. He was confident that it was a fallacy, and he suggested that for my course project, I try to unravel how infant water striders could propel themselves. Hunting down water striders to understand their motion didn't sound like research. It sounded like fun.

Seeing them in person was quite different from reading about them in a book. I tagged along with a group from my dormitory that was traveling to Walden Pond for a picnic. I used nets to hunt and catch the water striders. I kept them on top of damp paper towels, in the sandwich boxes that I had brought my lunch in. When I returned to John's lab, I dropped the striders into an aquarium full of water. They fell like feathers onto the water surface. As they stood on the water surface they used their long legs to groom themselves, rubbing as if playing the violin.

In the glass aquarium, I could see their legs from below (Plate 3). They looked like rowboats with six oars on the water. When the oars rested on the water surface, they were coated in a silvery layer of air, reflecting light like a diamond. If you were to submerge your own hand underwater, such a sparkling air layer would be absent. The water strider traps such a layer because it has the hairiest legs in the world, legs with 10,000 hairs per square millimeter, a million times denser than the hairs on a human head. Moreover, each of these hairs is covered in an array of grooves that make it even more water-repellent. The water strider's hairs, or trichia, were first pointed out by French and Danish naturalists in the 1950s. In 2004, Chinese chemists built a synthetic water strider leg by depositing fatty chemicals onto a smooth quartz fiber. Their synthetic leg could support the force of a water strider standing, but not the larger forces of gliding or jumping. The hairs on the water strider's legs thus have an important function in augmenting its water-repellency.

Although the functioning of the ornate details of the water strider leg remain under contention, one thing is clear: these hairs keep the strider dry by increasing the surface area of the leg. This is an interesting feature of surfaces. Using fractals or other types of patterns, one can increase the surface area of an object indefinitely. When such a large surface area is coated with a waxy water-repellent coating, water applied to the leg is kept at bay. Looking closely at the leg when it stands on the water surface, the water holds station at the tips of the hairs. It cannot penetrate the small grooves between the hairs. The result is that the water strider stands on water by standing on air.

The air layer on their legs gives water striders a remarkable ability, to glide on water for long distances, as if they are skating on ice. This ability has given the striders another name, pond skaters. During a rowing stroke, adult water striders clock up to 50 body lengths per second, the equivalent of a human running a 100-meter dash in a single second. After this high-speed stroke, they glide effortlessly for over ten body lengths. This gliding can be easily triggered: simply blow on a water strider. Like a puck on an air hockey table, the strider seems to just glide and glide. But if it glides so well, how does a water strider start moving in the first place? Where does the initial traction come from? The first person to recognize this problem was Stanford biologist Mark Denny, who wrote about it in his book on biomechanics in 1993. He came to the startling conclusion that he could not explain how water striders propelled themselves.

Mark proposed that the water surface changes state depending on how it is pushed. When the water strider glides, the surface is smooth like a linoleum floor. But when it rows backward, the water surface ruffles up like a trampoline, resisting the leg's motion. This resistance, the generation of a wave behind the strider, is enough to push the strider forward, so Mark thought.

Mark's idea of wave-based propulsion may work for the larger water-walkers, but for infant water striders it was problematic. Hydrodynamic theory states that surface waves can only be generated if an animal can move its legs fast enough. For straight-line motion at constant speed, that minimum speed is 23 centimeters per second, a low enough speed that it can be tested in the bathtub or swimming pool. If you dip a finger into the water and push it like a small boat, you do not create waves unless you travel faster than this minimum speed. While adult water striders have legs a centimeter long, infant water striders are ten times smaller, and their legs are barely one millimeter long. For infant water striders to achieve the required speed, they have to rotate their legs at rates higher than 1000 cycles per second, or 500 times faster than we can pedal on a bike. An infant strider trying to achieve these speeds would likely injure itself. The question of how infant water striders could move became known as Denny's Paradox, named after Mark Denny.

Denny's Paradox, and the idea that waves were necessary for water striders to move, spread across scientific zip codes. In Poughkeepsie, New York, Vassar biologist Robert Suter studied water spiders, thick-legged cousins of the striders. Suter observed that spiders also generated waves as they moved on the water surface. Even an isolated leg of a water spider, suspended in a flowing water tunnel, generated waves in its wake. Mathematicians from Stanford University calculated the wave field generated by a water strider's legs. The idea that waves were necessary for motion on the water surface became increasingly accepted by both mathematicians and biologists. But the more scientists accepted Denny's wave theory, the more puzzling Denny's Paradox became.

I had taken some photos of water striders for my project in John Bush's class, but I still didn't have an answer to Denny's Paradox. Resolving the paradox would require teamwork and some high-tech equipment so we could see more clearly what the leg was doing. I teamed up with my classmate Brian Chan, a mechanical engineer who was eager to build a robotic water strider. Satisfied with my performance in his course, John gave me the combination to the door to his basement lab, which was called the MIT Applied Mathematics Lab, a place where mathematicians did experiments with different kinds of liquids. It was filled with cameras and shelves of strange-shaped transparent vessels through which liquids could flow. At once, I had a PhD advisor and my own place to do research. My education in fluid mechanics had begun.

With the water striders now in the lab, we borrowed a high-speed camera from the Edgerton Center across campus. Our first high-speed videos of water striders showed quite a different world than what we could see with our eyes. The entire rowing stroke took a hundredth of a second. You could fit 30 rowing strokes into a single eye blink. With the camera, we saw that water striders rowed their middle legs with such force that their body was rocketed forward and upward into the air. Consistent with Mark Denny's predictions, the water surface behind ruffled up like a trampoline.

Denny's Paradox told us that the ruffled trampoline we saw could not be the only force driving water striders forward. The water striders were clearly getting thrust somewhere else, but where? We turned to *flow visualization*, the use of particles or dyes to pinpoint the motion of fluid. It's a technique that dates back hundreds of years, when visualizations of flying pigeons were made by throwing sawdust up into the air. As pigeons flew through, the falling sawdust spun into vortices. The strategy is to keep most of the fluid transparent, and place tracer particles in the region of interest. For the water striders, we needed a technique that stayed close to the action, the water surface. John knew of a technique used to visualize the impact of vortices using the chemical thymol blue. Thymol blue is a pH-activated dye, originally intended for chemistry experiments to show when a reaction was complete. We had not heard of it being used with live animals before, but why not?

In the lab, we began using the thymol blue to visualize the flows generated by the water strider's leg stroke. We first made the water alkaline by dissolving sodium hydroxide pellets, which look like rock sugar. We placed water striders on the water surface, and lit them from below using the kind of light box children use with tracing paper. Then we sprinkled in the dye with our fingers, like dried oregano. Sometimes a few flakes fell on the striders and they brushed themselves off, as if they were flicking off falling snowflakes. When the flakes landed on the water surface, the water blossomed into a deep blue color with small hints of yellow. When the striders were blown gently, they rowed, and the dye began to trace out their wake (Fig. 1.2, Plate 4).

With the aid of the thymol blue, we saw that the waves generated were only a brief manifestation of the rowing stroke. The waves came and were gone. Instead, what attracted our eyes was a butterfly-shaped dipolar vortex behind each leg. In general, a vortex is a spinning region of fluid, similar to the one generated in a cup of coffee when you stir it with a spoon. A dipolar vortex is generated from the rowing motion of an oar and has the shape of a palmier cookie. Even infant water striders were capable of generating vortices. The presence of such vortices provided a resolution to Denny's Paradox. By measuring their size and speed, we found they contained as much momentum as the water strider. The forward movement of the strider, like that of a fish, relies on pushing a packet of fluid backward.

Animals that propel themselves by pushing fluid backward must satisfy *conservation of momentum*. Consider a hummingbird that flaps its wings so that it hovers in midair. To hold station, it must constantly throw air downward. If the bird stops doing so, it sinks. Since air is so

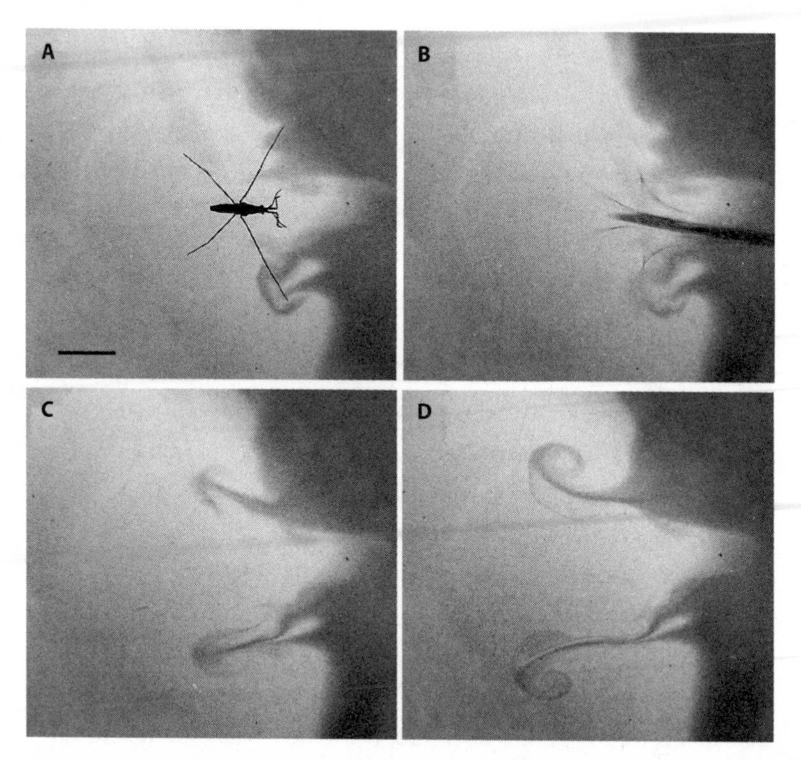

FIGURE 1.2. Vortex generation by the water strider. The strider rows its legs across a section of dyed fluid. The rowing stroke generates a pair of jets that roll up into a pair of dyed dipolar vortices. Seconds after their production, the hemispherical vortices are slowed to a halt by the effects of viscosity.

light, weighing 1000 times less than water, the bird must throw air at quite a high rate. The air that is thrown down has a certain momentum, the product of its mass and speed. The rate that momentum is thrown downward must be equal to the bird's weight for it to hover. A helicopter works the same way: its rotors push on air, speeding it up, increasing its momentum, and then throw the air downward as frequently as they can.

For a fish to swim forward, it must also satisfy conservation of momentum. As the fish moves forward, it flicks its fins and tail to generate an approximately fish-sized wake traveling in the opposite direction. Conservation of momentum dictates that this wake travels in the opposite direction of the fish. The form of the wake often depends on the mode of locomotion used. A tail fin generates a series of linked vortices like links of a chain. Birds flying at low speeds create a vortex with each flap, generating a series of vortex rings, like smoke rings. The wake of a water-running basilisk lizard consists of vortices that travel both downward to support its weight, and backward to enable it to generate thrust. The generation of vortices is a hallmark of the motion of many animals in water. Water striders live atop water, not in it, and it is likely for that reason that no one until John Bush had suspected they could generate vortices. Despite their sitting at the surface of water, water striders join the ranks of animals like birds, fish, and lizards that all create vortices to propel themselves.

If you have seen a water strider, you might be surprised that it can create vortices as large as watermelon seeds. A water strider is like a rower who has chopped off the blades of its oars and simply uses two thin sticks to row. In this case, its legs are fifty times thinner in diameter than the width of the vortex. How can such legs move the fluid? The answer is surface tension, but it arises in a subtle way. As a water strider sits on the water surface, it creates dimples, deformations of the water surface. These dimples look like small magnifying lenses centered on the strider's legs tips. The dimples are retained as the strider rows its legs. Filled with air but held together by surface tension, the dimples are like paddles that can be used to catch and push on more water than the insect could with just its spindly legs alone. Thus, water striders use their legs as shafts, their dimples as blades. The explanation was so elegant that we could hardly believe it. We had to try it out ourselves.

As autumn stretched into winter, water striders could no longer be found outside, but my classmate Brian had become a bit obsessed with them. One day, as we stopped to chat in the hallway, he pulled out the beaten notebook he carried around in his jacket pocket and showed me his drawings. They were a mix of mathematical equations and sketches of water striders of all shapes and sizes on the water surface.

The hardest part about building a robotic water strider is making it lightweight enough to balance on water. It's a tough constraint far beyond the capability of most robots. One of the most recognizable humanoid robots is Honda's walking robot called Asimo. It is four feet tall and weighs 120 pounds. Steel is heavy, and batteries, composed of metal, lead, and liquid chemicals, even heavier. In contrast, a typical water strider weighs a hundredth of a paperclip.

There was a small glimmer of hope. We had read about the largest water strider in the world, *Gigantometra gigas,* the giant pond skater, photographed only a handful of times in the rain forests of China and Vietnam, and never before raised in captivity. It looks like a normal water strider, except for its size: its body is three times as long as a normal water strider's, and its legs span nearly a foot. It weighed one gram, the weight of a single paperclip. While a typical water strider was too light for Brian to mimic, the giant water strider lay just within reach.

Brian was sipping a can of soda while we were discussing what material to use to build the mechanical water strider. We had learned from engineering class that one of the lightest and most affordable metals out there is aluminum, the same material used for soda cans. In fact, years of fine-tuning to cut costs have turned the soda can into one of the strongest objects for its weight. While its walls are only a tenth of a millimeter thick, as thin as a human hair, its walls are also strong enough for you to carefully stand on an empty soda can without its collapsing. The can maintains its shape until you tap the side wall with your finger, which causes it to instantly collapse due to *Euler buckling,* which is a danger for thin-walled structures. Soda cans are a miracle of engineering, but at the end of the day they are considered trash. Not to Brian. He finished the last bit of his soda, rinsed out the can in water, and walked over to the machine shop.

Brian worked in a machine shop like he was at home making a sandwich. He knew where everything was, and moved in a leisurely manner. He first cut off the ends of the can with scissors and flattened the rest into a thin shiny rectangle. He cut off a section and used pliers to bend it into a U-beam to strengthen it. This was Robostrider's body. He drilled four small holes at opposite ends of the body, where he would insert stainless steel wires and bend them into supporting legs. He cut two more lengths of wires for the driving legs, but these would wait until after the rest of Robostrider could float.

Cornell mechanical engineer Andy Ruina once said that the robustness of a robot can be measured by the distance between it and its maker. When we first placed it on water, we were like helicopter parents. Although the device weighed only 0.3 grams, the weight of one-third of a paperclip, it had to be placed very slowly on the water surface. Any quick movements would cause Robostrider to penetrate the surface and immediately sink. While Brian swept up the metal shards in his working area, I made a small testing arena that was lit from below so we could see the dimples clearly in the water. Brian placed the body of Robostrider on the water surface and there it floated, swaying from side to side, blown by the gentle breeze in the room. It reminded me of elementary school, where we learned to float paperclips. Now it was time to put a motor onto this high-tech paperclip.

Brian walked over to the lathe, a rocket-engine-looking device that turns rods into screws the way you might skin a carrot for dinner. He placed hard plastic rods into the lathe's gripper and locked them in. The rods were made of the same material used for guitar picks and zippers. After a few minutes of moving the lathe's cutting tool back and forth, he turned the machine off. The rod had turned into a pulley about the size of a chocolate chip. He inserted a long steel wire into the pulley and threaded it through Robostrider's head. The wire was then bent into two oars that stuck out from Robostrider like airplane wings (Fig. 1.3).

FIGURE 1.3. The mechanical water strider Robostrider faces its natural counterpart. Robostrider is 9 cm long, weighs 0.35 g, and has proportions consistent with an adult water strider. Its legs, composed of stainless steel wire, are hydrophobic, and its body is fashioned from lightweight aluminum. Robostrider is powered by an elastic thread running the length of its body and coupled to its driving legs through a pulley.

The power source for the water strider was the most difficult part to design. It had to be lightweight. More weight in the hull of an airplane requires longer wings to generate lift. Similarly, the heavier the water strider, the longer the legs had to be to support its weight on water. Most engineers would have given up at this point, and initially Brian was stumped.

When he was putting on his clothes the next morning, Brian had a eureka moment: he would use one of his athletic socks to power Robostrider. Today's athletic socks are a modern marvel. They are engineered to be as light as possible, yet hold onto a pair of legs without slipping down. They accomplish these tasks through a precisely engineered polymer-cotton blend, fabricated after years of collaboration between textile engineers, chemists, and material scientists. They were exactly what we needed. Brian picked a sock apart with tweezers, harvesting a small thread from the sock's elastic band. The elastic band was tied to Robostrider's rear and then carefully wound around the plastic pulley that drove its middle legs. The concept was simple, combining a simple windup toy with a small lightweight figurine of a water strider.

After Brian had assembled Robostrider, he carefully spun the legs to wind it up, and then kept his fingers on the legs to keep them locked. He carefully put it on the water surface. He held his breath, looked at me, and then released it. We saw a few water ripples, a blur of movement. Robostrider had gently coasted forward one body length, yet remained floating on the water surface. We cheered, and started gathering recording equipment ready to document its next walk on water. Just like for filming the water striders, we needed a high-speed camera. The human eye takes one-third of a second to blink, and we filmed at rates of 1000 frames per second, 300 times faster than a human blink. With the high-speed camera, we saw that Robostrider indeed was walking without breaking the surface. It was the world's first dry rowboat, able to move along without getting wet.

Over the years since Robostrider's first step, a menagerie of increasingly sophisticated robots has been built to walk, run, and even jump on water. Such robots are built with advances in micro-fabrication techniques, using lasers and light to make microscopic patterns into flat materials. Flat sheets of metal can be etched with small grooves to enable them to fold up like a pop-up book. One robotic water strider can even jump on the water surface to the same height as a water strider, 10 body lengths, or the equivalent of a three-story building for a human. The water strider's ability to jump on the water surface without breaking has to do with its flexible hairlike legs. These legs are flexible so that they can slow down the rate that force is imparted to the water surface. Again, like the thin cross section of the water strider's oar, this is a very nonintuitive idea. Our oars have always been rigid, with wide blades to catch the water surface. The water strider's solution is the opposite: to use thin flexible oars that do not break the water surface. This behavior enables the surface to stay intact and support the water strider's weight. The ability of evolution to arrive at solutions to problems that are nonintuitive yet effective will be one of the recurring themes of this book.

Today, the study of water striders and other insects has drawn the attention of computational fluid mechanics experts. These scientists use computers to predict the motion and forces of fluids. The water striders present a difficult problem for them because the boundaries represented by the water surface move and deform over time. One has to keep track of the pressures in the fluid at every point as well as the shape of the water surface. The numerous hairs on the water strider's legs also present modeling difficulties.

Water striders are just one of a number of water-repellent organisms that have captured the attention of chemists and materials scientists. Other organisms use variations on a similar theme: providing a rough surface that repels water. In the years preceding our 2003 publication on water striders, such water-repellent organisms created quite a stir. In 1997, German botanists Wilhelm Barthlott and Christoph Neinhuis discovered that the lotus leaf is covered in small round bumps covered in wax; when dirt and debris sit on top of the bumps, raindrops pick them up like a lint brush, rolling them away and leaving the surface clean. This ability is called self-cleaning, and has made the lotus leaf a religious symbol of purity for Buddhists. Water-repellent or "superhydrophobic" paints have been developed that have small beads mimicking the bumps of the lotus plant, and oils that mimic the wax. Such paints can be applied to car windshields to prevent water drops from sticking, but instead allow them to roll away like marbles. The problem is that the paints only have temporary effects due to the insidious nature of *fouling*, the deposition of contaminants from the environment. Another problem is that these paints can be chipped away by physical damage. A more permanent solution to these problems was suggested in 2012 by MIT mechanical engineers Dave Smith and Kripa Varanasi, who impregnate bumpy surfaces with edible oils that allow highly viscous liquids like ketchup to flow instead of sticking to the insides of bottles.

Truly permanent solutions to water repellency that can weather the harsh conditions outdoors are still over the horizon. To deter public urination, water-repellent paints have been applied on the outsides of buildings, and they repel urine so well they are known as "the paint that pees back." Nevertheless, such coatings have to be re-applied regularly due to the effects of fouling and physical damage.

Strangely enough, fouling does not seem to plague the water striders. When dew and fog roll in, they threaten to create small drops between the hairs of the leg that could make it lose its water repellency. However, on a regular basis, the strider rubs its legs together like a cricket. This causes the hairs to ratchet up and down each other, bending the hairs, which then shoot the drops out like catapults. Thus, not only are the hairs water-repellent, but they can be used to clean off water drops. Such self-cleaning water-repellent surfaces could solve the problem of fouling.

While I was studying water striders, I gave a lab tour to one of John's friends, a mathematician named Mike Shelley who supervised another Applied Mathematics Lab, this one at the Courant Institute of Mathematical Sciences, an independent division of New York University. Mike had just built a mathematical model for the motion of *C. elegans*, small worms that undulate their bodies to move. Such worms are used to study human health conditions such as aging and Alzheimer's.

He said that his mathematical model could probably be used for larger animals too. Mike and I talked about how a large class of animals, from snakes to worms, moved in this fashion. Like the water striders, they are ubiquitous, yet it was not easy to explain how they moved. I was intrigued, and with an invitation from Mike, I headed off to New York City to understand the motion of snakes.

CHAPTER 2

Swimming under Sand

An eight-year-old girl kicked her feet back and forth on the seat of a Long Island Railroad train. I beckoned her to come over and pointed to the top of my winter jacket, which I slowly unzipped. Inside, nestling against me for warmth, were ten snakes, their forked tongues waving back and forth. The child shrieked and ran back over to her mother, who was napping. "That man has a coat full of snakes," she shouted. Her mother grunted without opening her eyes, telling the girl to go back to sleep. The train's heater had not yet kicked in, and I was worried that the snakes would freeze to death without my body heat. I learned later from my biologist friends that they would have been fine in the cold, just uncomfortable. The snakes were my guide into the world of the sinuous, the world of long and flexible animals. It was 2006, and earlier that year, I had graduated with my PhD from MIT and I came to New York University to study the locomotion of snakes. I lived in an NYU-owned studio apartment near Greenwich Village, and was surrounded by a number of shops, but few places to buy snakes affordably. To start my experiments, I took the train up to a reptile expo held in a middle-school gymnasium on Long Island, where snakes that had been raised from eggs in basements in the suburbs were sold for cash only, no returns. There, I met New York City's "rat man," who drove around in an unmarked van, delivering freshly frozen bags of rats to the nearly one thousand snake owners throughout Manhattan. My snakes were not hungry yet, but I knew that soon after I returned to Manhattan, they would be.

For the entire train ride back, the snakes gave me an all-over massage. When I returned to my studio, I carefully removed all the snakes and laid them side by side on my bed like socks. They were so shiny they appeared to be wet, but were quite dry to the touch, despite the common myth that snakes are "slimy." I spent a few minutes just admiring how beautiful the snakes looked up close. The corn snakes were brown with white and black speckles, like kernels on an ear of corn. They were docile, and moved slowly, more willing to wrap around my finger than to escape. The garter snakes were green and black, with texturing like a garden hose. They were more skittish, and moved their bodies away jerkily, releasing musk if spooked. I had also brought home two boa constrictors. One of them was a six-foot-long, twenty-pound snake named Houdini. Some days later, Houdini broke out of his glass terrarium and simply disappeared. I searched my entire studio apartment, but figured that he had crawled out the open window and into a sewer. A month later, as I was getting dressed, I found Houdini hibernating in my underwear drawer. Houdini and the rest of my snakes represented only a portion of the range of snakes in nature: a thread snake can fit on a quarter, while a reticulated python can extend 30 feet, as long as a three-story house is tall. More than 2000 species of snakes inhabit the forests, deserts, and nearly every continent around the world. I was eager to see them all, but for now, I was content with three species inhabiting my apartment.

I had seen snakes in magazines before, but I had never held one in my hands. I discovered that holding a snake was a sensuous experience. It wrapped its body around my wrist, hand, and fingers, constantly adjusting its position so it would not fall. It was mesmerizing to watch,

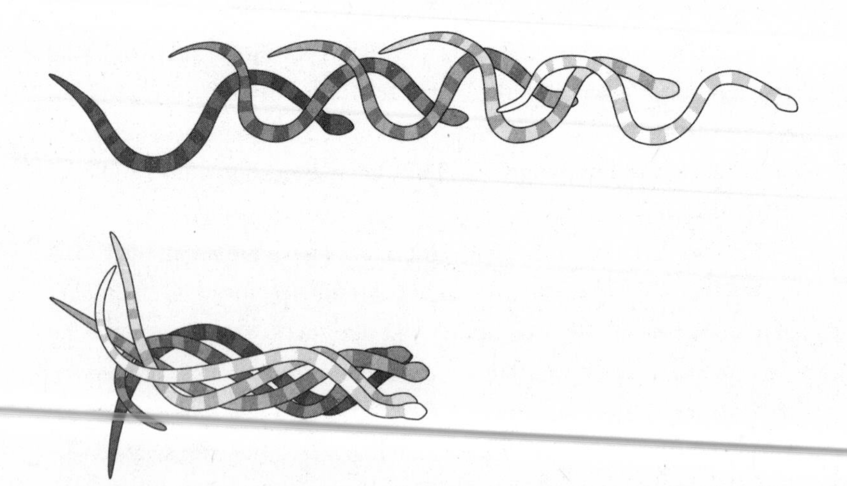

FIGURE 2.1. Multiple exposures of a snake slithering on two different surfaces: carpet (top) and hardwood floor (bottom). On both surfaces, the snake undulates its body with the same frequency and wavelength. However, only on the rough surface does the snake generate forward motion.

and difficult to predict which direction it would travel. Animals like cats and dogs can run quickly, and their inertia requires them to exert effort when they want to make a turn. In contrast, snakes don't need brakes. They can start and stop on a dime. A snake seems to apply force with its entire body, like one continuous leg that it can wrap around anything it wants. It is quite different from the way we move about.

The earliest work on snake locomotion was published in the early 1930s. Dutch, American, and British biologists, including Walter Mosauer and James Gray, proposed that snakes pushed off the rocks in their surroundings, the rocks providing push-points for the snake's flanks. By pushing on one rock after another, a snake slithered forward, each part of its body slaloming like a skier between poles. The idea seemed reasonable, except that there were many environments devoid of push-points. Granite, flat rock, deserts, and asphalt were all relatively featureless. What would the snakes push on when they got there?

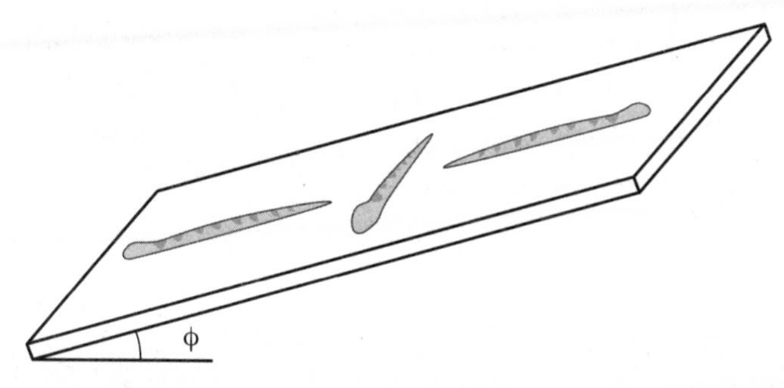

FIGURE 2.2. The inclined plane apparatus used to measure a snake's friction coefficient, or resistance to sliding. We temporarily put the snake to sleep and place it in various orientations on an inclined plane. By slowly angling the plane, we record the angle at which the snake begins to slide.

To understand how snakes could move without push-points, I chased them all over my small apartment. I noticed the snakes slithering easily across carpeted surfaces, and then flailed about on smooth ones (Fig. 2.1). Legend has it that the Taj Mahal was designed with smooth marble floors because snakes could not easily slither across them. I wondered what it was about the carpet that allowed them to slither. But the snakes were moving too fast for me to see.

I wanted the snakes to stand still, so I used the knockout gas isoflurane to get them to take a short nap. A sleeping snake is like a limp rope, very different from the clingy creature that crawls continuously upward in your hands. Once the snake was motionless, except for its body gently rising and falling, I began the experiments to measure the snake's resistance to sliding. I found an old bulletin board in the lab, and used thumbtacks to attach to it a large piece of green cotton fabric that I would perform all my experiments on. I gently laid my snake atop of the fabric. Then I slowly raised one end of the board, turning it into an inclined plane (Fig. 2.2). If I raised the board slowly enough, at some critical angle, say five degrees, the snake would simply slide down the ramp, like an avalanche of snow sliding down a car windshield. This experiment sounds quite simple, but it yields useful information: the angle of the ramp at which the snake slid can be written mathematically into a term called the *coefficient of static friction*, a number that describes the snake's inherent resistance to sliding. The more I could incline the ramp without the snake slipping, the greater the snake's frictional resistance. It was important that I did this experiment while the snakes were sleeping. Awake snakes can sense the motion of the ramp and open and close their scales like you would draw venetian blinds, delaying their sliding until the ramp is nearly 40 degrees. That is how snakes can climb tree trunks. I liked this experiment because it was very robust: the numbers extracted were quite repeatable. The first time I did an experiment like this was in elementary school, where I learned that all materials, from glass to wood, each have their own friction coefficient. The only difference was, back then, I didn't have to worry about my test subjects waking up.

When the snake slithers, it creates an S-shape curve, and slithers its body in various directions (Fig. 2.3). When it pushes backward, the ground pushes the snake forward. The higher the friction coefficient, the more force is applied to the snake. Snake locomotion is then just keeping track of each of the directions that the snake is pushing itself. Imagine that the entire snake is like an ice skater's blade. To predict what direction the snake will go, I have to measure its resistance to sliding in each direction, which I can use to calculate the total thrust force on the snake. I measured the snake's resistance to sliding by placing it in different orientations on the ramp, and tilting until the snake began to slip. I found the snake had the least resistance when it slid down the ramp head first: in this orientation, I could barely lift the ramp five degrees before the snake began sliding. In contrast, when the snake was placed backward or sideways, it was more resistant to sliding: for the latter, I had to lift the ramp over 10 degrees. This dependence of friction on the direction of motion is called frictional *anisotropy*. It helps snakes to climb uphill, and it arises from the arrangement of their belly

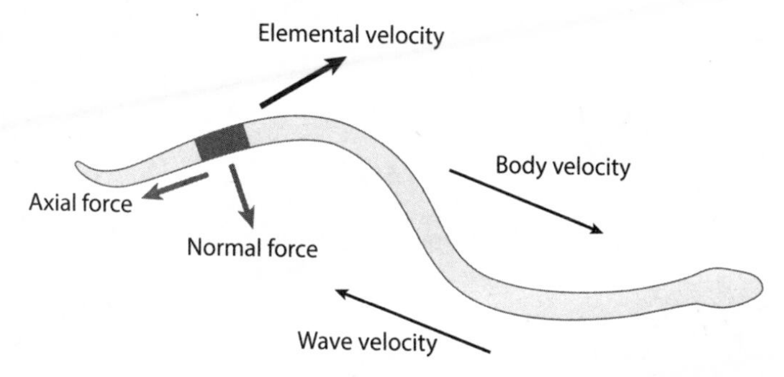

FIGURE 2.3. When a snake undulates its body, it generates reaction forces that act on small elements, shown by the shaded box. These forces can be divided into a normal force that acts perpendicular to the element, and an axial force that is parallel. The sum of the forward normal forces must exceed the backward axial forces in order to propel the entire body forward. In other words, by pushing laterally a snake can overcome the friction of dragging its body.

scales, which you can see if you flip a snake over (Fig. 2.4). A corn snake's belly scales are arranged in a sawtooth pattern, like overlapping venetian blinds. When I ran my hand in one direction they felt smooth, but in the opposite direction, my hand caught on the edges of the scales. By pushing toward its flanks as it slithered, the snake could turn this resistance to sliding into thrust, the same way we push ourselves forward when we walk: we need high friction on our shoes to get traction as we push off the ground.

Using Mike's mathematical model for worms, I combined the slithering motion I recorded on video with the friction of the snakes' scales. Mike wrote computer code that calculated the motion of the snakes, then showed on-screen what that motion looked like. When the computerized snake had friction that was constant in every direction, as if it were on linoleum floor, it slithered in place. Moreover, the computerized snake moved forward only if it had frictional anisotropy, as if it were on a carpet or other textured surface. All this matched up with my experiments.

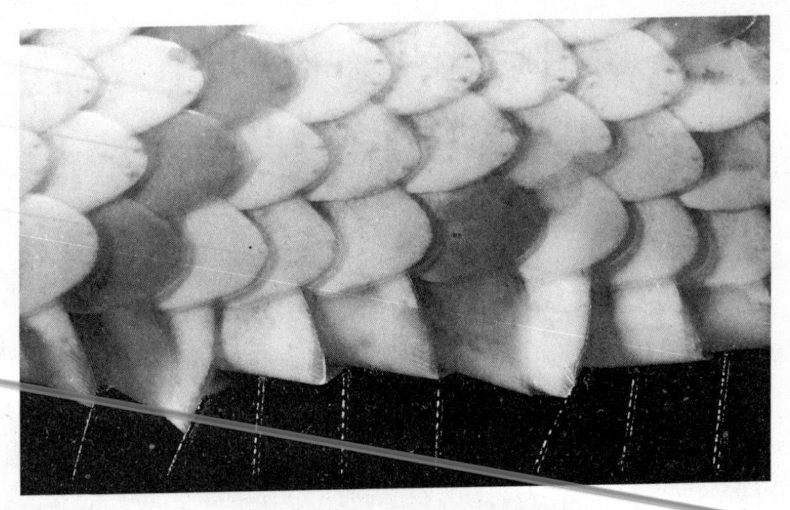

FIGURE 2.4. The wide overlapping belly scales of a corn snake interacting with an array of plastic pillars. The snake's scales have an important function: they snag on ground asperities, which gives the snake a preferred direction of sliding on surfaces of sufficient roughness and compliance.

While Mike Shelley and I were celebrating, we also noticed a slight problem. The computerized snake only moved half as fast as its natural counterpart. Somehow, the scales that I had assumed were responsible for facilitating locomotion were not the whole story. I returned to the snakes in my apartment and let them slither around the floor. I watched them over and over from various angles. The day drew on and I started drooping, finally conking out on the ground in exhaustion. I lay there watching the snakes, and the surprising number of dust bunnies I had under the couch. As a snake slithered past me, I noticed that I could see small flashes of light shining underneath its belly. I knew that there were gaits like sidewinding where snakes lifted entire portions of their belly like a corkscrew. But I had thought that such body lifting was limited to sidewinding. Later, I would do experiments with snakes on gelatin, a material that becomes a pressure sensor under polarized light (Fig. 2.5). These gelatin experiments would show that a

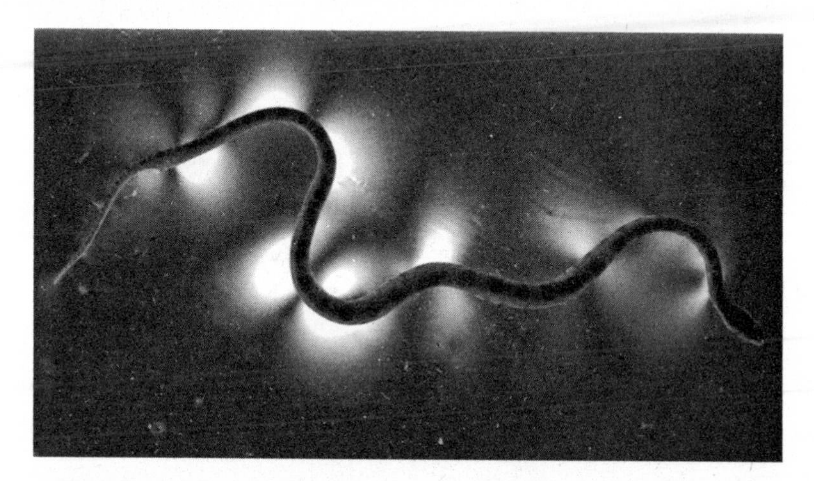

FIGURE 2.5. A corn snake slithers across photoelastic gelatin. The long molecular chains in gelatin are stretched when the snake pushes, enabling polarized light to pass through. Luminescent areas indicate regions of strongest applied force, showing that the snake applies forces non-uniformly along the ground. Rather than pressing itself uniformly along the ground, the snake lifts sections of its body while slithering to increase speed and fuel economy.

slithering snake can adjust its weight distribution even if parts of body are lifted just a sliver above the ground. But for now, my mind was racing because I realized that in Mike's computer model, I had it all wrong: I had assumed that the snake pressed its entire body evenly on the ground as it slithered.

I rushed downstairs and across the street to the Courant Institute and told Mike about the idea excitedly. After a few trials on the computer, I found that with some minor adjustments in weight distribution, I could recover the majority of the missing speed. The snake was doing the same thing as when we walk, lifting our leading foot rather than letting it drag on the ground. I found that by simply lifting the peaks and troughs of the snake's curves, I could double its speed and reduce the energy spent in locomotion. The surprising effect of lifting is due to the dual role of friction in snake locomotion. Friction has two sides: it can either slow down a snake or speed it up, depending on what direction that part of the body is going. Specifically, the peaks and troughs of the snake's body do not provide thrust force. Because they are primarily pushing laterally, they only accrue drag, slowing the snake down. If the snake wishes to reduce drag, it needs to lift these body parts off the ground. If it does so, the only remaining points of contact with the ground are the snake's inflection points, the center of its S-shaped curves. These are the only parts of the snake that are being pushed backward while the snake slithers forward. Thus, they act like the driving foot of an ice skater. By applying more of its body weight at its inflection points, the snake pushes itself forward as an ice skater would by pushing down on her driving leg. Thus, to travel forward, the snake must send its body waves backward. The analogy with ice skating goes even further: if I switched the direction of slithering for the computerized snake, I could get it to travel backward, just like an ice skater can. The same friction coefficients can lead to forward or backward motion, depending on the direction of the traveling wave employed.

* * *

The US military employs Humvees and tanks, vehicles that touch the ground with large wheels or treads, as large an area as possible to avoid sinking. Nevertheless, the contact area is small compared to the size of the vehicle. In contrast, a limbless animal propels itself using its entire surface area along its long, sinuous body. This large contact surface enables it to travel to places that wheeled vehicles cannot go. Snakes are found on nearly all the continents on earth. They can swim and climb trees—some can even glide from tree to tree. Other limbless animals such as worms are responsible for aerating and turning over the world's mud and soil. Even plant roots have long, sinuous shapes that enable them to push through dirt and around underground rocks.

Roboticists are beginning to recognize the benefits of making long, sinuous robots. Snakelike robots climb trees to take video reconnaissance or to send signals. They can slither underneath doors or within pipes. Snake-inspired medical robots are also entering hard-to-reach places in the human body. Many organs in the human body are protected by a stiff rib cage. Open-heart surgery, for example, requires the rib cage to be sawed in the middle and cracked open by force. Minimally invasive or laparoscopic surgery is one alternative. Small holes are made in the body cavity; long, sinuous robot arms enter and, controlled by joysticks, make cuts and perform suturing. The recovery post-surgery is far faster with the use of these kinds of robots. Currently, such robots are being designed to provide tactile feedback to the user's joysticks, as the robot arms are pushed through skin and bump into organs inside the body.

We say that long, sinuous animals are *hyper-redundant* because they have a great number of repeating bones and muscles. A snake can have hundreds of vertebrae, whereas humans only have 33. The effect of this large number of vertebrae is that the snake can form long, sinuous curves, where every part of the body can touch the ground and generate forces. Such organization gives the animals tremendous versatility and multifunctionality. A snake can stand up on its tail, balancing to climb a tree. It can wrap its tail around a branch and cantilever its body to reach across gaps.

Our previous discussion concerned motion atop surfaces. However, animals can also move through flowing materials such as mud and sand. These animals are known as burrowers. They move materials differently than we do with our burrowing machines, which can move huge amounts of material, digging entire tunnels for trains and roads. Such high throughput is quite well engineered. By looking at burrowing animals, we can learn how to make small handheld burrowing machines, used for reconnaissance or to measure local properties of soil, reaching places that would be inaccessible to us above ground. The secret to effective burrowing in animals is using their entire body as a tool. As we will see in the next story about worms, a long, sinuous body can be both a shovel and a chisel, enabling the worm to find weaknesses in materials and exploit them.

* * *

It was low tide, and University of Maine doctoral student Kelly Dorgan stomped across the mudflats in the small fishing town of Edgecombe. She felt like a fly in pudding. With each step, she laboriously lifted up her boots, and the mud made sucking sounds. Her Black Labrador Rossi also moved with difficulty, her paws covered in mud. In the distance, a few figures were scattered along the shore. They were the wormers of Edgecombe, grizzled seamen with chins covered in white bristle, stooped over and hacking at the mud. The area around seemed otherwise deserted, but Kelly and the wormers knew better.

Beneath her feet, in the depths of the mud, lay an entire bustling city. In just one mile of shore lay hidden seven tons worth of marine worms, known as *polychaetes*, "those with many bristles." Polychaetes are the gardeners, the janitors, and the caretakers of the mudflats. They digest living remains and till the mud, aerating it for bacteria and other mud-dwellers. Unlike human farmers, polychaetes never take a holiday. They are the foundation for life in the mudflats, and are absolutely essentially for other life-forms. Without worms, as Darwin noted in his last book, *The Formation of Vegetable Mould through the Action of Worms,* the rest of the ecosystem would quickly fall apart. Despite their importance, the worms do their work in secret. When we pick up a rock, they quickly hide away.

We live in a world of light, where going from point A to point B is simply a matter of walking. We take pleasure in freely raising our arms or kicking our legs. Worms live in a completely opposite world. It's dark, damp, and surrounded on all sides by heavy mud. Despite these close quarters, worms must constantly move forward, seeking food, air, and water. But exactly how worms move through mud remained a mystery until Kelly Dorgan completed her doctoral thesis on the subject in 2007. Before that, biologists assumed that they just ate their way through the soil, much as a worm eats its way through an apple. Soil would simply come in through the mouth and exit through the anus. The problem with this picture is that there are many different types of soil, many of them very hard. How do worms burrow through such a wide variety of soil?

Ever since Kelly was in elementary school, she was fascinated by slugs, bugs, and all sorts of slimy things. She spent her childhood in Yorktown, Virginia, catching crayfish, playing with them, and sometimes then keeping them as pets. A high school teacher introduced her to marine worms, and she was hooked. She was drawn to them for their abundance, which, to her, indicated their importance relative to other animals. But for nearly everyone else, worms were unnoticed and misunderstood. She enrolled as a college student at University of California, Santa Cruz, where she learned of the diversity of invertebrate life in the ocean, and was particularly attracted to marine worms (Plate 5). Marine worms are soft, unprotected, and bright red, free lunch for a seabird if they do not remain hidden. To find them, Kelly had a few tricks the seabirds did not have.

As she walked across the mudflats, Kelly carried a bucket in one hand and a garden pitchfork in the other. Once she got to the right spot, she stooped down and began hacking at the mud with her pitchfork, her movements fast just like the wormers had taught her. The mud was fine. It had the texture of gelatin and came apart easily. Once she saw the red tail of a worm, she grabbed it quickly before it wiggled itself into the mud. The worm was as long as her index finger and covered in small soft spikes, as if it were wearing a felt Halloween costume. It wriggled around quickly as she tossed it into her bucket. Rossi barked behind her in support, looking for worms too, but mostly just getting in the way.

In half an hour, Kelly had collected 37 worms. She started walking back to the lab, where she had several tanks of synthetic mud already made. Formulating the recipe for this mud was a task that had taken her weeks. The problem with natural mud was its opacity. To see the worms, Kelly had to create transparent mud that had the same properties as real mud. Making mud pies when she was younger, she knew that mud could be made at a wide range of consistencies depending on the water content and the sediment type. Chefs have taken advantage of this fact for years, creating foods that give the tongue a range of texture and softness. As a result, a number of food-grade gels were readily available online. These included gelatin and agar, as well as an assortment of bizarre-sounding but ubiquitous thickeners such as carrageenan, glucomannan, alginate, and xanthan gum, which are in most processed foods, if you look carefully at the packaging. By mixing this variety of gels, Kelly could imitate muds across a variety of locations and seasons. To measure their softness, she pushed on the muds with force sensors, as we would squeeze a peach for ripeness.

Kelly's gel mixtures fell into two broad classes, creamy muds and stiff muds—analogous to mud pies with lots of water, and little water. Creamy mud was like a freshly made mud pie. It was soft, and it deformed easily when she pushed on it. Stiff mud, on the other hand, was like a mud pie that had been dried in the sun. It easily cracked in two, each half crumbling in her hand like blue cheese.

Kelly started with the stiff mud. She had made a gallon, filling an aquarium the size of a shoebox. She used forceps to open a crack in the surface of the mud, and carefully placed one of the worms into the crack with tweezers. Before long, it began to dig its way in. It wiggled its head back and forth in the entrance of the burrow, which opened the crack of the burrow even more (Fig. 2.6). As the crack extended deeper, the worm could wiggle more and more of its body into the crack. Before long, its entire body was underground.

The worm's behavior was in stark contrast to how biologists had thought worms moved. In fact, the motion of the worm was strikingly expert. As Kelly would learn, the worm was exploiting the properties

FIGURE 2.6. The marine worm *Nereis virens* burrowing through gelatin, which simulates muddy sediment. The worm generates a crack-shaped burrow, which is easiest to extend at the crack's tip. Illumination using polarized light turns gelatin into a visualizer for forces: the bright regions in the photograph indicate the high stresses at the tip of the burrow. Courtesy of Kelly Dorgan.

of the mud to move forward using the least amount of energy. This method is called *propulsion by crack propagation.* We have yet to make a machine that does this so well. Imagine trying to cut into a cheesecake that you have previously frozen in the freezer. If you are like me, you will simply take a cleaver and pound and pound at it until it is broken in half, possibly breaking your countertop at the same time like I did once. The worm's strategy, on the other hand, is to make a little notch, and then just wiggle to push the sides of the notch apart. I tried it, and it works great with frozen cheesecake. But why is it so effective?

The worm's strategy takes advantage of a property of stiff materials, discovered during World War I. During the war, airplanes and tanks were built with high-strength materials such as steel and glass. The mystery was that in practice, these materials were not very strong, breaking at applied loads that were 100 times less than expected based on the strength of the molecular bonds. What caused this apparent weakness? In 1920, British aeronautical engineer Alan Arnold Griffith solved this puzzle using experiments with glass. He conducted experiments with smooth glass, which he could indent with a very small notch. Such notches are a controlled way to introduce a crack into glass. Griffith conducted fracture experiments with different notch sizes and made a striking observation: if the notch exceeded a certain length, later called the Griffin length, the glass was doomed to break, and at very low applied forces. Cracks cause failure because the tip of the crack is under very high stress. His experiments explained why glass broke at such low stresses. You can demonstrate the same principle by trying to tear a piece of writing paper apart by pulling on the two sides of the paper lengthwise. You'll find the paper holds together very well. But if you tear a small notch in the middle of the top edge of the paper, it tears apart quite easily. Griffith's discovery led airplane designers to abandon popular manufacturing techniques like cold rolling that introduce small cracks into the material. They instead turned to polishing their metals to remove cracks. The new increased strength of materials ultimately led to larger airplanes, such as the Boeing 727, that have a fully cantilevered wing made of a single sheet of metal. The days of using trusses in airplane wings were over.

A crack is the Achilles' heel of a material, and the worm takes advantage of it to move through a mud that is much stiffer than its body. It as if the worm is the head of an axe. Once there is a crack in a block of wood, it doesn't take much to break it apart. There are certain situations when taking advantage of crack propagation is difficult for the worm. For instance, consider when the worm is jammed between mud and an aquarium wall. When the worm wiggles its head, the rigid wall pushes back. As a result, the worm has to apply 10 times more radial force to move along the wall than when it does far away from the wall.

Next, Kelly turned to the creamy mud. The worm's method of generating a crack would not work here. Imagine trying to cut whipped cream by wiggling a knife. The material is too soft and will deform before it cracks. But the worm has a backup plan. It inflates its body as much as it can, and pushes against a small cavity that it makes in the soft mud. This motion anchors it. The worm pauses briefly, perhaps taking a deep breath. It pushes its head forward into the notch, and then blows a party horn out of its mouth. The party horn is actually the worm's throat, which it can turn inside out and extrude. The party horn parts the mud enough to make a crack grow in front of the worm. This method is similar to how earthworms push themselves through soil. Because the mud is so creamy, the crack is quite short and can barely be seen, unlike the long visible crack in stiff mud. Once the crack is formed, the worm moves forward and anchors itself again. Forward motion is slow and is measured in blows of the worm's pharynx.

Scientists have described over 15,000 species of polychaetes, and another 30,000 are estimated to exist. They are quite diverse: some swim, some burrow, and some even possess plantlike abilities to grow roots into hard materials. In 2002, the "bone-eating" polychaete, *Osedax*, was discovered in decaying whale bones nearly three kilometers below sea level. *Osedax* has no mouth and no anus, but feeds using a complex system of roots like a tree. The roots excrete acids that can penetrate bone, and bacteria living in the roots can digest the fat reserves found deep within the bone. *Osedax* and other worms play an important part in recycling carbon that falls to the ocean floor. Polychaetes turn this organic matter into carbon dioxide and nutrients, which eventually return to the surface so that small marine algae called phytoplankton can use it for photosynthesis. Without this critical link, the ecosystem would be broken. Unfortunately, the number of polychaete species is disappearing due to global warming and other human activities. Kelly is among only 200 scientists in the world who study polychaetes, aiming to decipher their secret world beneath the mud.

Kelly's polychaetes travel through a wet world of mud that can either crack or deform like cream. Outside the mudflats, soil is just as plentiful, but the moisture may not be. This is the case for the world's deserts, where a multitude of other animals live. While we would walk across the dunes, a number of animals have learned to travel in the cool beneath the sand.

* * *

"I heard you have a lizard that can swim under sand like it's water," said University of California, Berkeley postdoc Dan Goldman. He was standing in the doorway of the office of Ted Papenfuss, a herpetologist who worked at Berkeley's Museum of Vertebrate Zoology. Ted beckoned Dan to come in. On a trip to Iran in 2001, Ted had met up with Bedouins, Arabic nomads who lived in the desert. In one of their tents, they had offered him a mysterious delicacy called a sandfish, which they had cooked over coals. A live sandfish now resided in a glass aquarium lit by a heat lamp on Ted's desk. The tank was filled halfway with sand. It appeared otherwise empty except for a few insect legs and body parts, remains from the sandfish's last meal. Ted threw in a cricket from a nearby tank. The cricket walked about on the sand, sending vibrations underneath that the sandfish could detect. Without warning, a patch of sand near the cricket moved. A sandfish leaped out, grabbing the cricket in its jaws. The sandfish was not a fish at all! It was a yellow mottled skink, or lizard, about the length of a human hand (Fig. 2.7). It thrashed the cricket with shakes of its head to throw off the extra sand that it had also snatched. Then it crushed the cricket several times in its jaws, making loud snapping sounds. Dan was amazed. After they eat, Ted said, they simply disappear into sand, and are nearly impossible to catch. As soon as Ted reached in, the sandfish dived back into the sand as if it had just disappeared into thin air. He began combing his fingers through the sand. Suddenly, a patch of sand twitched. Ted blindly grabbed at the

FIGURE 2.7. (Left) The sandfish *Scincus scincus* uses its shovel-shaped nose to swim underneath sand. (Right) A high-speed X-ray image of a 10-centimeter-long sandfish reveals that to "swim" within sand, the animal does not use its limbs but instead propels itself using body undulation. Courtesy of Dan Goldman.

patch, and somehow was able to pull out a wiggling tail. At the end of the tail squirmed the sandfish, which he handed to Dan. Once held, the sandfish stopped struggling and turned to the side, cocking one eye at Dan.

Typical lizards have bodies shaped like a loaf of Italian bread, rounded on the flanks. The sandfish's body has four edges, and, like a bus, its sides are perfectly flat. As Dan would find out, these were adaptations to help it reduce forces when swimming under sand. In fact, when it breathes, its flanks do not expand like typical animals, but only its chest expands. Almost nothing was known about how it moves, Ted said. Dan was intrigued.

What drew Dan's attention was not the sandfish, but the sand it was burrowing through. To anyone else, sand might seem simple enough, but Dan knew that nothing could be more complex. Dan's graduate training was in granular physics at the University of Texas at Austin. He knew sand as one of many *granular materials*, materials composed of individual dry grains, sticks or other rigid objects, such as sand, snow, twigs, and even chicken nuggets. When sand flows, the sand grains smack into each other like billiard balls, their energies lost to the forces of friction. This behavior gives sand, in a sense, characteristics of all three phases of matter, gas, liquid, and solid.

Snow is one example of a granular material. It can cap a mountaintop, and without warning, it can suddenly avalanche. This is because the snow fluidizes at the boundary between the sliding pile and the mountain's edge. Once at the foot of the mountain, the snow solidifies again. Water cannot behave in such a variety of states. We would be shocked to see rivers flowing and then stopping and then flowing again. Nineteenth- and twentieth-century mechanics, a branch of physics, had been devoted to understanding solid materials that can bend and fracture, or liquids that can drain and flow. But much of twentieth-century mechanics was devoted to understanding complex materials, such as non-Newtonian fluids like mud and yogurt and granular materials like sand, all of which exhibit behaviors of multiple phases of matter.

As Dan left Ted's office, he replayed in his head the sandfish's dive into sand. Since the sandfish was so difficult to catch, he surmised that it could not only dive, but could also move beneath sand. How did the sandfish propel itself subsurface? Did it build tunnels to run through like a mole? Did it use its forelimbs to push its way through sand in a kind of breaststroke? Gaining evidence for any of these hypotheses was difficult because of the opacity of sand. We can't see through sand until we move it out of the way. But to move it out of the way would mean perturbing the system that we were trying to measure in the first place. Dan needed a way to see through sand without touching it.

Since the 1930s, the airplane industry had been trying to solve the same problem. They had airplanes undergoing several flights every day, experiencing large forces on the wings that would cause them to age. The entire airplane had to be regularly checked for damage, quickly yet thoroughly. In the 1980s, one of the tools used was X-rays to penetrate the wing to look for cracks and corrosion that were not visible from the outside. One of the companies to use this technique was Aircraft X-ray Laboratories, just 10 miles south of Berkeley. Dan asked Ted if he could borrow the sandfish, and took it down to the X-ray facilities.

Looking at an X-ray of a sandfish is like watching a ghost (Fig. 2.7). The sandfish body has disappeared, and all one see is a cavity in the sand. This is because of the nature of X-rays. X-rays are simply beams of light with enough energy to penetrate through solid objects. The denser the object, however, the less deeply the beams can penetrate. The sandfish is made mostly of water, and its surroundings mostly sand. Since sand is denser than water, the sand around the sandfish appears opaque, and all one sees in an X-ray video is an empty lizard-shaped cavity burrowing underneath the surface. Although Dan could see the location of the sandfish under X-ray, the images were not sufficiently high quality for Dan to distinguish what its legs were doing. Later as assistant professor at Georgia Tech, he built a higher-resolution version of the X-ray. He and his student Ryan Maladen purchased an old dental X-ray camera. They lined an entire basement room, walls and ceiling, with lead, to prevent the X-rays from leaking out. They wore lead vests and turned on the X-ray only for short bursts to avoid harming the sandfish.

Dan and Ryan built several new pieces of equipment, specially designed to test locomotion in sand. Having the proper equipment helps immensely to measure the effectiveness of locomotion. As I discuss in subsequent chapters, every animal has a particular "test bed" best suited to studying it: fish are filmed in water tunnels, birds in wind tunnels, legged animals on force platforms. All of these pieces of equipment have one thing in common: they make the particular conditions of that medium repeatable. Water and wind tunnels do this by making the flow uniform, creating a medium that acts like a blank slate. But how to create repeatable conditions in sand?

When Dan was a graduate student at UT Austin, his advisor Harry Swinney took him on a trip to see another professor who was an expert in fluidized beds. A fluidized bed is a device that makes sand bubble and float atop an upward stream of air. It was discovered by accident in 1922 by the German chemist Fritz Winkler, who was trying to expose coal briquettes to gases to convert them into a gaseous state that could be more easily transported to homes. Winkler's fluidized bed consisted of a bin of coal, under which was a porous filter. A fan sent air through the filter. At low speeds, the air would simply pass through the holes between the coal briquettes. But if the air was pumped at high enough speed, the aerodynamic forces on each piece of coal would be sufficient to levitate it. The resulting cloud of floating, colliding particles is called a fluidized state, and it can be created in any collection of granular materials. It was useful to Winkler because the fluidized state increased the contact of each of the briquettes with the incoming gas, which in turn increased reaction rates. Nowadays, the fluidized bed is used in the food industry as well, as cold air can facilitate the rapid freezing of food products such as peas and shrimp.

Years later, Dan recalled that the fluidized bed was just the right tool to study sandfish. With a few pulses of air, Dan could tune the level of packing of the particles, similar to the way you can shake a box of dry oatmeal flakes to get it to settle. One of the difficulties with the fluidized sand bed was that every time he turned it on, huge billows of dust were released like in a dust storm. Natural sand has a range of particle sizes, and the grains rub on each other to create more powderized sand. He could solve this problem by replacing the sand with a larger substitute. He had found that the sandfish could swim through anything that was made of grains. Poppy seeds, glass beads, any particles that were up to one centimeter in size would induce the sandfish to dive right in. For his experiments he settled on three-millimeter glass beads, similar in size to small couscous. These particles were small enough to permit the sandfish to behave naturally, but large enough that they did not generate dust.

A bullpen for the sandfish sat across an enticing bed of glass beads about a foot long and a few inches deep. When the door to the pen was opened, the sandfish burst out, and immediately dived. X-ray video showed that after the sandfish dived, it tucked its legs close to its body and swam like a crocodile. It swam quite quickly through the glass beads, traveling at nearly two body lengths per second, equivalent to a car traveling at 20 mph. The sandfish's speed was much higher than expected. Was that speed the result of the looseness of the medium? The beads in the first test were loosely packed, like the sand that has been blown onto a front porch. Dan was able to use the fluidized bed to change the packing state of the beads, making it more similar to the packing in a sand castle. As he increased the density of the beads, the X-ray videos showed that the sandfish maintained its high speed. For a long time, this result mystified Dan. Consider trying to move through a crowded train. As the train gets more and more packed, you will have to move slower because of the increasing number of collisions with those around you. The sandfish, on the other hand, could cruise at the same speed no matter how much debris was in its path. This strange result indicates the strange physics underlying motion through sand.

To understand motion in sand, let's first consider a tuna. A tuna moves through water by pushing water backward using swipes of its large tail fin. As it throws water backward, the fish is driven forward. The sandfish wiggles its entire body, in essence using its entire body as its tail. To understand how a wiggling body travels, we use an idea called *resistive force theory*.

Sand, unlike air or water, is an inertia-less regime, meaning that sand grains do not flow unless continually pushed upon. After you dive into a swimming pool, the water continues flowing, evidenced by ripples on the surface of the pool. Eventually, the ripples disappear and the water is still once more. The energy of the moving fluid, its kinetic energy, has been transferred into heat by the water molecules rubbing against one another. In contrast, after you jump into the sandbox, there is no wake. The sand becomes still almost instantly. This is because the grains of sand transfer energy into heat much more quickly. This is the key difference between swimming in sand and swimming in water. As each part of the sandfish pushes through the sand, the sand flows, but ceases with the cessation of the applied force. Thus, the forces on each part of the sandfish may be considered independently.

Imagine the sandfish as a sausage that we chop into a series of shorter segments. We wish to estimate the total force on the sandfish as it wiggles each of these segments back and forth in a wave. To measure the force on a single segment, Dan used a robotic arm to drag a metal rod through a bed of glass beads. This metal rod represents one segment of the sandfish's body. The rod was pushed in one of two principal directions: *parallel* to the rod's axis like a thrown spear, or *perpendicular* to its axis like a rowing oar. Each of these principal directions is associated with different forces. Driving the rod in the spear direction generates drag due to friction with the sand. This friction slows the sandfish down. Rowing like an oar generates thrust. Each part of the body acts like an oar or like a spear, as the wave travels through that part of the body. In more densely packed sand, the sandfish experiences greater drag because the grains are more tightly packed around it. This would suggest that the sandfish would slow down. But because the sand is so tightly packed, each oar-like motion of the sandfish also gives it more thrust. The higher thrust balances out the higher drag, leading the sandfish to travel at the same speed using the same body wave motion. Keep in mind, however, that the sandfish does not maintain its speed for free: it must expend more energy to produce the same body wave motion in the denser medium. If you'd like to imagine what it is like for the sandfish, consider a similar experiment done by Ed Cussler and his graduate student at University of Minnesota. They managed to convince the swimming pool staff to fill the pool with xanthan gum, a thickening agent that increases the water's viscosity by a factor of two. Despite the water being more viscous, swimmers were able to achieve the same swimming speed. Like the sandfish, they were likely expending more energy to do so.

Underneath sand, the sandfish rocketed itself along at two body lengths per second. Its high speed through a dense medium like sand began to inspire scientists and engineers to think of machines that could also move quickly within or across natural terrain. At the time of Dan's sandfish work in 2009, much research with robots had been on laboratory linoleum floors, or on asphalt. Terrain in nature is much more challenging. It is covered in grass, leaves, debris. Its height has small deviations and is sometimes soft, preventing wheels from making full contact. Consequently, robots get stuck, sometimes spinning and digging themselves deeper. In 2012, the $400 million Mars rover Spirit suffered the same problem. Spirit had a rectangular base with six large grooved wheels. Powered by a solar panel, it was a trouper: envisioned to last 90 days on Mars, it managed to survive six years. The rover's ultimate demise was not a lack of power but simply a patch of loose sand in which it became ensnared. For nine months, it attempted in vain to escape before being retasked as a stationary research platform.

Saving wheeled vehicles from flowing ground like mud and sand is an ancient problem. The Egyptians pushed large stone blocks on an array of logs to distribute the blocks' weight. In 1877, the Russian inventor Fyodor Blinov invented the tank tread, at the time called the "wagon that moved on endless wheels." The treads help to spread the weight of the vehicle across a larger surface area, preventing individual wheels from sinking. The idea works well if the ground is sufficiently flat. Since then, treads have been incorporated into tractors, construction vehicles, power shovels, and tanks. Increasing the level of "aggression" of the track, also known as the sharpness of its points, be they metal or rubber, helps to increase traction and prevent sinking over mud, soil, snow, and other yielding surfaces. The main problem with tank treads is that the treads and the machinery to run them are heavy. The added weight requires more fuel and a stronger motor to push, which again adds weight, requiring yet a stronger motor, and a wider tread. Consequently, treaded vehicles are often heavy and power-hungry.

That changed in 2001 with University of Michigan electrical engineer Daniel Koditschek, who led a multi-university team to build RHex, the first fast lightweight robot to move on natural ground. *RHex* stands for running hexapod, a name for six-legged insects. It looked like a toaster with six C-shaped legs, with no eyes or other apparent sensors. But this robot could run! Its strategy, which was different from many robots, was to simplify. Although it had six legs, it did not move them independently. Instead, it grouped them into two sets of three. Consider the front and hind leg on the left side and the middle leg on the right side, all lifting in sync, as they are in Fig. 2.8. This tripod of legs lifts in the air first, while the remaining three legs keep their position. Then the roles reverse. The leg tripods thus alternate, giving this gait the name *alternating tripod gait.* This gait is the most common gait in the world, the default gait for ants, cockroaches, and most six-legged insects; if we were insects, this would be what we call a walk. Controlling the robot legs in this way reduced the on-board computations and the weight of the robot even further. The robot ran in "open loop," where it did not receive feedback from its environment, but simply rotated its legs. Surprisingly, this strategy worked great, allowing the robot to travel at two body lengths per second, which is the equivalent of a person taking a jog at 8 mph. Imagine doing that with your eyes closed. That's what RHex did.

Dan Koditschek designed a version of RHex called Sandbot to specialize in traversing sand (Fig. 2.8). There is a lot of sand in places

FIGURE 2.8. The robot Sandbot, a small version of RHex, a series of biologically inspired robots that move on six legs. Sandbot moves using an alternating tripod gait in which two sets of three C-shaped legs rotate synchronously and out of phase. It runs on a bed of poppy seeds that challenges forward motion, causing Sandbot to slowly sink into the ground, like a car in mud. Adjustments of the motion of the legs can restore good performance by utilizing the liquid-solid behavior of the poppy seeds. Courtesy of Georgia Tech.

where autonomous robots like Sandbot were needed, such as the war zones of the Middle East and the dunes of Mars—places too dangerous to bring a human passenger but that Sandbot could easily explore with a camera and a radio antenna to send back images. To test how well Sandbot would work on sand, Koditschek sent it through the mail to Dan Goldman.

Although Sandbot ran quickly on hard ground, the first time Sandbot touched sand, it was a disaster. Its speed reduced by a factor of 10, from a jog to a tiptoe. And as it continued walking, it sank deeper and deeper until it was completely stuck. Moreover, after a few runs, the sand got stuck in Sandbot's fragile gears. Imagine a grain of sand stuck in your eye. When that happens, you know to stop everything and lift your eyelid gently to allow the grain to fall out. Sandbot doesn't have those fine sensations. The grains just get caught in its gears until the gears are fully jammed; then the motor overheats, and must be replaced. Sand is made from silica, one of the most common materials on earth, but also one of the hardest. When gears face off against sand, sand wins.

Two problems had to be solved to get Sandbot to run on sand. First was creating a safe training ground for the robot. This was done by replacing the hard sand with the soft toppings from a bagel. Poppy seeds. They were two millimeters in diameter, nearly 10 times larger than sand grains, but the robot was already so much bigger than a grain of sand that the size difference did not matter. Poppy seeds that became entrapped in Sandbot's gears were so brittle that they simply crushed into powder. Now that the robot was safe, Dan could start looking at the second and more difficult problem: at why exactly it was getting stuck. This is where an understanding of the foot's interaction with the sand became important.

Typically, Sandbot rotates its legs at constant speed, about five rotations per second. Every time a leg strikes hard ground, the ground simply pushes back on the robot, moving the robot forward. When watching the videos of Sandbot on the poppy seeds, however, it was clear that there was little push-back. Sandbot's legs struck the ground so hard that poppy seeds flew everywhere. The escaping seeds prevented the robot from getting much thrust, and without the seeds to support the robot's weight, it sank deeper and deeper with each step. To get the robot to work, the researchers had to slow it down so it could take advantage of the natural properties of sand.

Consider the properties of surface tension. With gentle leg motions, a water strider can stand and walk on water as if it were atop the skin of a pudding. But if the leg were to break through the surface, it would need to move much more quickly to generate the same thrust. Sand, in a sense, exhibits similar behavior. At low applied forces, sand is a solid, and supports a foot as if the sand were solid ground. At high forces, it yields and flows like a liquid. The robot's strategy, then, would be to avoid fluidizing the sand, which meant fast leg movements when the leg was in the air, but then slow movements when it contacted the sand. Slow leg motions would obtain more resistance from the sand, and more sure footing than fast leg motions, which would fluidize the sand. The concept is similar to driving a car in mud: if you floor the gas you'll only spin your tires, while if you give the car just a bit of gas, you can slowly drive out of the mud. The solution worked and the robot was able to travel across sand. The strategy showed that by simply changing the leg's motion, the robot was able to recapture its performance. The key to motion on sand was to treat sand gently to avoid fluidizing it.

* * *

The snake, worm, and sandfish are all examples of animals that appear to swim on land. To propel themselves, they take advantage of the properties of the medium, its friction, its ability to crack, and its tendency to fluidize. Much of what enabled the animals in this chapter to succeed was their particular body shapes—in this case, long and sinuous. In our next chapter, we will examine further repercussions of shape.

CHAPTER 3

The Shape of a Flying Snake

As I accepted my award, I shook hands with Dudley Herschbach, Nobel laureate in Chemistry. Seven other Nobel laureates stood nearby, beaming their approval. I made my way to the Harvard University podium to address the audience of more than a thousand. Having reached this high point in my career, I noticed a few things were different than I had imagined. The toilet seat I wore around my neck was heavy, and began to slide. Behind me stood the human spotlight, a naked middle-aged man painted in silver paint, one of the inventors of full-color 3D printing. As I walked, I felt the crunch of paper airplanes littering the stage, left from the ceremony's grand opening, when thousands of paper airplanes were thrown toward a human target who wore a lab coat, safety glasses, and flashing red lights. When I arrived at the podium, a red-headed eight-year-old girl with freckles, pigtails, and overalls stood next to me, her arms crossed. If my speech went even a second over time, she would start yelling "Please stop, I'm bored" repeatedly until I stopped. This was not the Nobel Prizes. It was the Ig Nobel, and I couldn't have been happier.

Just a year earlier, I was in a much darker place. My son Harry had been born and my wife and I were in triage mode. My wife took charge of feeding, and I took charge of the other end of metabolism, diaper changing. Harry was rambunctious and treated everything like a game. When it came to diaper-changing time, he would try to crawl away, hiding behind the couch, giggling and shrieking. Eventually, I would catch him, covered in dust bunnies, and bring him to the changing table as he flailed his arms and legs. As I removed his diaper, he giggled even louder. It was just another day in fatherhood until his urine stream hit me squarely in the chest.

This was a first for me, and I could feel the anger rising in my chest. I heard my wife's voice in my head, telling me to count out loud to calm myself. One, two, three, I counted. I kept counting, and the urine kept coming. Gradually, the jet lowered to a mere trickle, and I stopped counting at 21 seconds. That is a long urination, I thought, as I put on his diaper. Perhaps a bit too long.

For a ten-pound kid, Harry sure had a big bladder. Urine is made in our kidneys, which filters urea from our blood. Therefore, urine volume should be proportional to the blood volume of the body. But my body weight was at least ten times my son's. I should have ten times as much blood, and in turn, ten times as much urine volume. Why, then, is his urination time so long? I began to worry. Perhaps my son had a serious medical problem, like some obstruction that made him urinate for longer times. I started imagining the waiting room in the pediatrician's office, filled with screaming kids.

I put my son down on the floor and headed to the bathroom to clean myself off. When I saw the toilet, I decided to do my own experiment. I pulled down my pants, put one hand against the wall, and started counting. It was the most important pee of my life. One, two, three . . . I counted up to 23 seconds. Wow, I thought, my son urinates like a real man already! I should have named him Hercules. How could my son be urinating for the same time as me, despite his having ten times less urine volume?

The answer to this question would drive me into thinking about the shape of the urinary system, and more broadly about the importance of animal shape in driving the flow of fluids through and around animals. As I began to think about animal shape, I started to see animal motion as more than just about getting from point A to point B. Animals use motion of their body parts, internal and external, to perform a variety of functions, including cleaning, grooming, eating, and digestion. All these tasks involve moving liquids and solids from inside to outside the body, and vice versa.

Animals surround us with a seemingly endless variety of shapes. We see far more variety in animals than we do, say, in sports cars, which are all generally streamlined to reduce drag. In part, this is because there is more to shape than just getting from A to B the fastest way—otherwise, every animal would be aerodynamically streamlined like a sports car. Underwater and in air, there are different forces at play. For urination, the bladder and urine are subject to gravity. Underwater, gravity doesn't matter because of Archimedes' law. Since water-dwelling animals are the same density as their surroundings, the pressure of the surrounding water supports their weight. As a result, a huge variety of animal shapes are possible underwater, from sharks to rays to jellyfish. For flying animals, top speed and fuel efficiency may not always drive evolution. Instead, robustness to stall or robustness against incoming air currents may be the drivers. Because of the different environments, and the different needs of each animal, a number of shapes arise, both inside and outside animals' bodies. This chapter relates three stories in which people have become obsessed with animal shapes. Keep in mind that these shapes are not the best by any means, but are what has emerged through evolution as good enough.

* * *

The day my son urinated on my chest, I was also scheduled to teach undergraduate fluid mechanics. I thought it was the perfect opportunity to tell the class about my experience. I asked the class if they had any explanation for the long urination time. The class seemed mystified and I saw a few whispering to one another. I asked the students to raise their hands if they wanted to try to help answer this question using science. A premedical student—who would later become a urologist—and his friend volunteered, and after class I ushered them into my office. For this experiment, I had exactly the right tools for them to test their mettle. I gave them a few stopwatches and a mud-splattered bucket that I used to gather ants. I told them, please bring these items to the Atlanta Zoo, and don't come back until you've measured the urination time of every animal there.

Attacking the zoo with a bucket and stopwatch can't be done without having the proper training. To train the students, I recruited Patricia Yang, a tall, lighthearted Taiwanese graduate student who had studied physics and ocean engineering. She and the undergraduates practiced catching urine from dogs at the local park. This task was more difficult that I originally anticipated. My initial idea was to follow dogs with plastic cups, trying to catch their urine. It was one of my lab's most failed experiments. Every time we saw a dog urinating, we would all sprint toward it like a football team trying to catch a ball. Of course the dog would see us and panic, stopping urine in midstream and running in the opposite direction, growling at us. With this method, we didn't get a single drop of urine. Over time we improved our methods. We used dog training pads to catch urine. Then we would weigh the urine in the pad, using its density, which is similar to water, to calculate the volume.

Patricia and her student charges practiced similar procedures at local farms, measuring the urine duration and volume of goats, sheep, and cows. Now that they were fully trained, it was time to tackle the zoo animals, which had to stay in their enclosures to keep everyone safe. Therefore, they had to measure urination time in a much less invasive manner. Every afternoon for a period of a few weeks, they simply stood outside the animals' enclosures in the hot sun, armed with a large reflector to shine light at the animals' genitals, my high-speed camera, and plenty of patience. Plate 6 shows the urination stream of an elephant, when their urine-filming abilities were at their peak. Every time an animal urinated, they wrote down how long it took.

After a few weeks of hard work, they came back from the zoo. They were dust-covered, urine-spattered, and looking rather disappointed. I asked them what was wrong, and they said that they had worked very hard, but had not found a very interesting result. They had measured the time of over 40 animals, but found that their urination time did not vary much from animal to animal. More than 70 percent of the time, urination was between 10 and 30 seconds long, with an average of 21 seconds.

Well, that is the most interesting of all, I exclaimed. We looked at the data more closely and my eyes ran down the variety of animals they had studied, including dogs, goats, pandas, rhinos, and elephants. We did not control for diet, gender, or time of day. The experiment was entirely uncontrolled, and yet, the animals had a strikingly consistent urination time.

While some would argue that 10 to 30 seconds is a wide range, one must keep in mind the widely varying bladders involved. My dog Jerry had a bladder volume of about a cup. An elephant has over 100 times that volume, able to fill up an entire 20-liter kitchen garbage can. You would expect the elephant to take 100 times as long as Jerry, or at least 10 times as long. Yet they both emptied their bladders in the same amount of time—at most the difference was a factor of two. At conferences, I would ask fellow scientists how long an elephant took to urinate. Most of them gave me a dirty look or just ignored my question. When I pursued this further, the answer I could get out of them was usually, about a minute. No one would have guessed that an elephant and my son urinated for the same time.

My students and I pored over anatomical diagrams of the urinary systems of hundreds of mammals (Fig. 3.1). While mammals are all

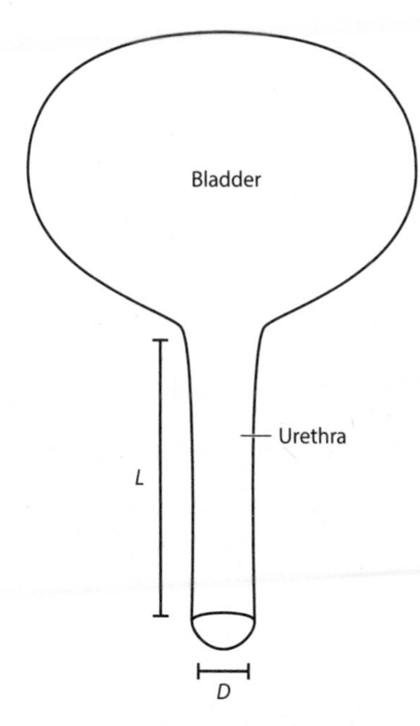

FIGURE 3.1. Schematic of the urinary system of a mammal. The bladder contains urine, and a pipe underneath called the urethra transports the urine out of the bladder. In mammals of the same sex, the pipe has a constant ratio of length to diameter. Gravity causes fluid to drain from the bladder.

shapes and sizes, we found that evolution was remarkably consistent with the urinary system. For all mammals, the urinary system begins with a balloon called the bladder, which stores the urine. Beneath the bladder is a long pipe that physicians call the urethra, and I call the pee-pee pipe. My son Harry likes to tell his sister that only he has a pee-pee pipe, but I have to remind him that he is mistaken. Both men and women have urethras, with a ratio of length to diameter of 25 to 1 and 17 to 1, respectively. Moreover, these ratios are conserved not just across humans, but also across all mammals, from mice to elephants. Despite the urethra's ubiquity, at the time we conducted our study, no scientists knew why mammals needed urethras in the first place. In fact, most other aspects of the mammals change radically with body size. Elephants have larger ears to cool themselves and thicker legs, proportionally, than smaller animals. Both are a consequence of their large size. An elephant needs thicker legs to support its massive body weight. If most body parts change in proportion with increasing body size, why does the urethra stay conserved? My hunch was that it had to do with the function of the urethra: to release fluid from the bladder.

To understand how fluid flows out of the urethra, we have to consider the concept of pressure. It is a nonintuitive and subtle concept, one that mystified the ancient Greeks. Their top philosophers described a number of hydrostatic paradoxes, situations where liquids poured into static containers behaved in unexplainable ways. This confusion lasted all the way to 1646, when, according to legend, the French mathematician Blaise Pascal used to stun audiences with an experiment that became known as Pascal's Barrel. Legend has it that he would tell the audience that he could use fluid to easily break a wine barrel, one of the sturdiest structures around at the time. He then climbed up a ladder and attached a very thin, 10-meter-long metal pipe to the top of the barrel. A hole at the top of the barrel allowed the contents of the pipe to flow directly into the already full barrel. He began to pour water down the pipe. Since the pipe was so thin, the fluid in the pipe was far less than that in the barrel. But as the fluid topped the pipe, the seams of the barrel cracked. With a loud snap, the top of the barrel blew off and the water began leaking out. I've done this same experiment in my fluid mechanics classes with plastic lunch Tupperware, a pressure gauge, and a long rubber pipe. It really works.

Pascal's demonstration shows that a fluid's pressure has nothing to do with the amount of fluid involved. Instead, it matters how the fluid is configured in space. This is a very different idea than we are used to with solids. If we have a lump of clay, we don't think that the clay exerts different pressure depending on how it is shaped. Yet, this is how fluids behave, because they can transmit forces differently than solid materials. Pascal later showed that the pressure in the pipe is proportional to its height, a fact now known as Pascal's Law. He could thus increase the height of the pipe to make the pressure in the barrel arbitrarily high, regardless of how thin the pipe was. In this way, fluids can amplify the force of gravity far more than solid materials can.

The urethra is simply the biological version of Pascal's Barrel. By making the urethra taller, animals can use gravity to increase the pressure driving the urine. An adult woman's urethra is five centimeters long and the width of a coffee stirrer. But a female elephant's urethra is one meter long and the width of your fist. These dimensions enable the elephant to urinate with a flow rate of five showerheads. The larger flow speed and the wider area of the pipe compensates for the larger volume in the bladder, enabling large animals to urinate in the same time as smaller animals. My graduate student Patricia derived the mathematical formula for urination time, which we called the Law of Urination. She found that if the aspect ratio of the pipes remains the same, the volume released does not depend on the volume in the bladder; it simply remains at 21 seconds. Such principles of designing pipes to speed up flow can be used in designs of water towers, water backpacks, or even apartment buildings. They show that through clever design of pipe area and length, fluids can be tuned to be excreted in constant time, regardless of the volume.

How did animals evolve a constant proportion of urethra? I don't have an answer to this question, but I speculate that the constancy of urethra shape was likely a process of "good enough" prevailing. I imagine that a short urination time was driven by the threat of predation. However, 21 seconds appears to be a good enough value to reduce this threat sufficiently. A shorter urination time would require a urethra so long that it becomes unwieldy or a urethra so wide that parasites could more easily enter. These evolutionary pressures may have helped maintain the urethra aspect ratio to be constant. To further understand the evolution of the urethra, it would be interesting to examine the urethra of whales and other mammals that left land for the open oceans.

The urethra is just one of many bodily organs whose shape is critical to its function. Other organs in the body—the eyes, lungs, heart—often change in shape with body size. The study of this change is a subject called *allometry*, a word coined by English biologist Julian Huxley in 1932, meaning the growth of body parts at different rates resulting in a change of body proportions. Allometry is communicated using mathematical expressions called *allometric scalings*, which relate an animal's mass and the dimensions of the body part such as the length, width, or volume. If you imagine animals lined up in a row, from smallest to largest, these scalings give a rule for predicting the shape of the body part as the animal changes in size. The urethra has a special kind of allometric scaling because its proportions do not change with body size, a property called *isometry*. Isometry can be easily observed in the toy industry. Dolls and toy trucks are recognizable because they have the same proportions as the items they are modeled on.

From your own experience, you might observe biological shapes often do not follow isometry. For example, an infant's head is a much bigger proportion of its body than an adult's. As a child grows, its head grows at a different rate than its body. Such shape changes can be used to infer the physical constraints on these organs. In 1637, Galileo was one of the first to recognize that bigger animals have different proportions than smaller animals, a feature most prominent in animals' leg bones. When scaled to the same size, an elephant's femur appears short and squat compared to a dog's femur. This difference in proportion permits the elephant to support its larger weight. Later, engineer Thomas McMahon extended Galileo's initial observation. He measured a number of museum specimens and found leg bones could be described by an allometric relationship or mathematical formula: the leg bone diameter is proportional to the length to the 3/2 power. This exponent causes larger animals to have relatively thicker bones. For example, an elephant's femur bone is ten times as long as a dog's but twenty times as wide. Bones that satisfy this scaling are called "elastically similar" because this shape helps the bones maintain constant stress, and so resist Euler buckling. In recent years, an increasing number of bones were found to not satisfy elastic similarity. Brazilian biologist Guilherme Garcia has shown that bone shape is also influenced by factors such as growth and the force applied by the muscles, which in turn will change with locomotion type. A number of theoretical models exist, but it still remains a mystery what factors set the allometry of bird and reptile leg bones.

Allometry is a tool to measure the shapes not just of bones, but of soft body parts as well. In our next story, we will learn how the shape of jellyfish can lead to inferences about how optimal their stroke is.

* * *

On a sunny afternoon on the San Juan Islands, off the coast of Seattle, a team of scientists led by Caltech professor John Dabiri waded along the rocky seashore, surrounded by jellyfish. The jellyfish were in the thousands, appearing just under the surface of the water as ghostly silhouettes. Since his days in graduate school, John had been thinking about jellyfish motion. Wading along with him were his teachers, Rhode Island biologists and jellyfish experts Sean Colin and Jack Costello, who were longtime experts at working with the soft, fragile bodies of jellyfish. John had also brought his graduate student Kakani Katija, with whom he had studied jellyfish for a number of years. John's teammates also served an important purpose on this trip: John couldn't swim. Over the last few years, John had been working with the US Navy on a new type of submarine. He had a hunch that the jellyfish was ejecting fluid in a way that would be useful for the submarine to emulate. By studying the jellyfish, John would bring new ideas to the Navy on their quest for a new, improved submarine.

At the time, the state-of-the-art for driving submarines was the propeller, or what the Navy calls the screw. The screw is the heart of the submarine, and it works because it drives the submarine quietly. A loud submarine is a found submarine. The loudness is generally due to the generation of small bubbles, which come about through *cavitation*, the process of pushing water so fast that it vaporizes. Years of research into the propeller's particular shape have reduced cavitation and helped the propeller push the envelope for speed. The idea for the propeller itself is quite old. It was originally invented by Archimedes in 200 BCE for irrigation. The mathematical theory for the ideal propeller was developed in 1865 by William Rankine, one of the founding fathers of thermodynamics. Since then, the propeller has been propelling submarines underwater.

John's idea was to build a submarine that pushed itself like a jellyfish or squid. These animals propel themselves by jetting. They have a cavity that fills with fluid, which through their contraction of their flexible bodies pushes fluid backward like a water balloon. It's underwater rocket propulsion, albeit using a slow and steady rocket. Since he was in college, John had seen jellyfish in the aquarium and was fascinated by them. They were such fragile-looking animals, yet they seemed to move so effortlessly. John wanted to integrate their mechanism into his submarine. His idea was to open and close inlets to the submarine, mimicking pulses of a jellyfish. Inside the submarine, a screw would continuously push out fluid. The timing of the opening of the inlets would set the length of the jet produced. Before he embarked on making prototypes of his submarine, he decided to observe the jellyfish in their natural environment. To obtain the biggest diversity of jellyfish, John and his team had traveled up the coast from Los Angeles to Seattle, where he knew jellyfish would be teeming at the shore.

The water gently lapped at John's knees, sunlight glistening off the water surface. He had his trousers rolled up and was dressed for collection, wearing a T-shirt, and a single black rubber bracelet on which he had printed a biblical verse to remind himself that he was only one person in a large universe. As the jellyfish lazily rose to the water surface, John and his team bent down to collect them. Nets would damage their fragile bodies, which cannot resist the slightest forces. A netted jellyfish pulled ashore would look like a puddle of motionless goo. Instead, the team had brought glass baby-food jars, which they used to directly scoop the jellyfish and the water surrounding them. Each jar would fit a few jellyfish, each the size of a thumbnail. They gathered the jellyfish quickly, yet carefully, so as not to damage them.

To the team, collecting jellyfish was like gathering seashells. Each one was a little bit different (Fig. 3.2). *Neotourris* was a "fast," bullet-shaped jellyfish. *Aequorea* was one of the more leisurely ones, shaped like a dinner plate. Some shapes seemed even ostentatious—*Leuckartiara* looked as if a carrot and a pear had a baby. At the time of John's fieldwork, no one understood why the types of jellyfish were each shaped so differently. Jellyfish also seemed like the least likely place to look for inspiration for submarine design. But the assortment of size and shape was exactly what John wanted. He wanted enough jellyfish that he could have a foot race between them, and then use the winner for his submarine design.

When John and his team finished filling his jars with jellyfish, they waded back to the rocky shore. The sharp smell of pine trees wafted over from the forest beyond the shore. They reached the dock, where several boats with nets were moored. Ahead, a small group of white buildings looked like a fishing village. They formed Friday Harbor Marine Labs, a fully stocked laboratory for studying marine animals and plants.

John and his team brought the jellyfish to a lab room that they had rented. The room was dark except for a green sheet of laser light that shined into an aquarium filled with seawater. Small bits of plankton and other natural debris whirled about, illuminated by the laser sheet. Using the laser light, they captured not only the motion of the jellyfish,

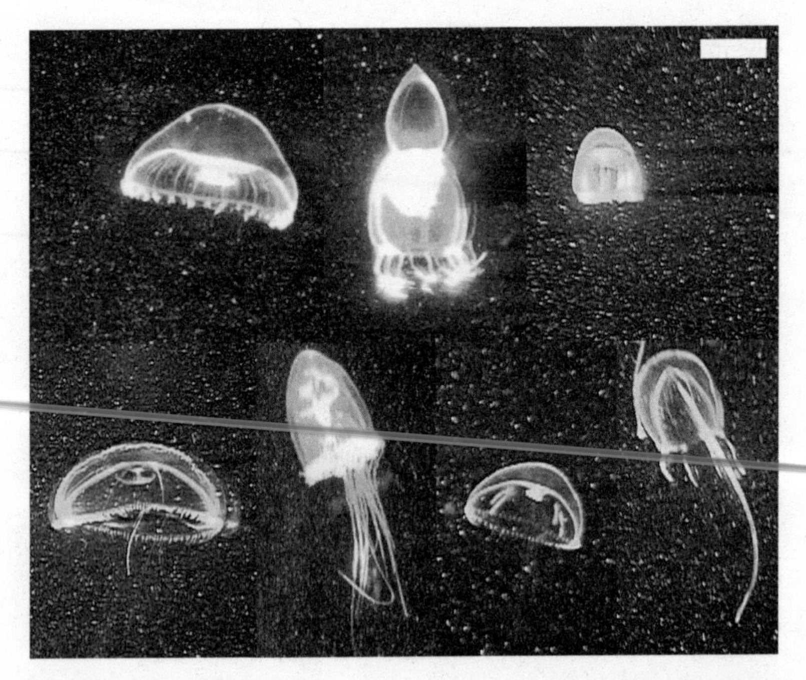

FIGURE 3.2. Jellyfish propelling themselves by contraction of their bell-shaped bodies. Jellyfish speed is strongly affected by body shape and the resulting shape of the jet produced. In particular, tall jellyfish propel themselves faster, squat ones with lower energy expenditure. A laser illuminates the jellyfish bodies as well as natural particles in the seawater. (Top row) *Aequorea, Leuckartiara, Melicertum,* (Bottom row) *Mitrocoma, Neoturris, Phialidium, Sarsia* sp. Courtesy of John Dabiri.

but also the flow of the water that was jetted from the jellies. The room felt almost like a dance club, with the glow of the laser and the low hum of the high-speed camera's small internal fan.

Taking turns filming, Kakani, Sean, Jack, and John eventually obtained films of over 50 jellyfish from seven species. The jellyfish naturally split into two groups based on shape: the dinner-plate jellyfishes and the bullet-shaped ones. The shapes correlated with their swimming speed and their way of life. For instance, the dinner plates were plankton eaters. Their food just floated around waiting to be eaten, so the jellyfish could take their time getting there. The bullets had to be fast to chase small fish. This much was known about the jellyfish. However, what was not understood was how their shapes influenced their speed. Intuitively, the profile of their bodies was suggestive. The bullets cut through the water easier than the dinner plates. This was the drag part of the story. But the thrust part was subtler, and was related to John's original questions on submarine design.

How did body shape affect the flow generated? John analyzed the motion of the fluid behind both the dinner plates and the bullets. The bullets created a wake that is similar to the one we create when we swim: a vortex followed by a messy trailing jet, full of eddies (Fig. 3.3C). The wake looks like the jet of a water fountain. The dinner plates, on the other hand, created a much cleaner wake. It consisted of a single vortex ring, similar to a smoke ring blown from a smoker's mouth (Fig. 3.3A). The rest of the fluid was quiescent. Imagine swimming through a fluid and leaving most of the fluid motionless except for a single swirling vortex. Upon analyzing the momentum in each, he found the dinner plates were much more economical with their energy, using far less energy for each centimeter traveled. These jellyfish ate plankton, a low-calorie source of food, making saving energy more important than speed. Conversely, the bullet jellyfish pumped much more energy into each stroke. They were exceeding the energy they could put into a single vortex ring, and as a result the ring went unstable, leaving behind a messy wake. They did so because going fast was much more important than saving energy. If their prey escaped, all their effort would be wasted.

John found that jellyfish body shapes dramatically affected their swimming ability, separating them into sprinters and distance runners. He lined up the jellyfish in order of aspect ratio. Consider the jellyfish like soda cans, each with a ratio of its length to its width. The long skinny bullet-shaped ones would be at the top, then the squarish ones, and lastly the flat ones resembling dinner plates. For reference,

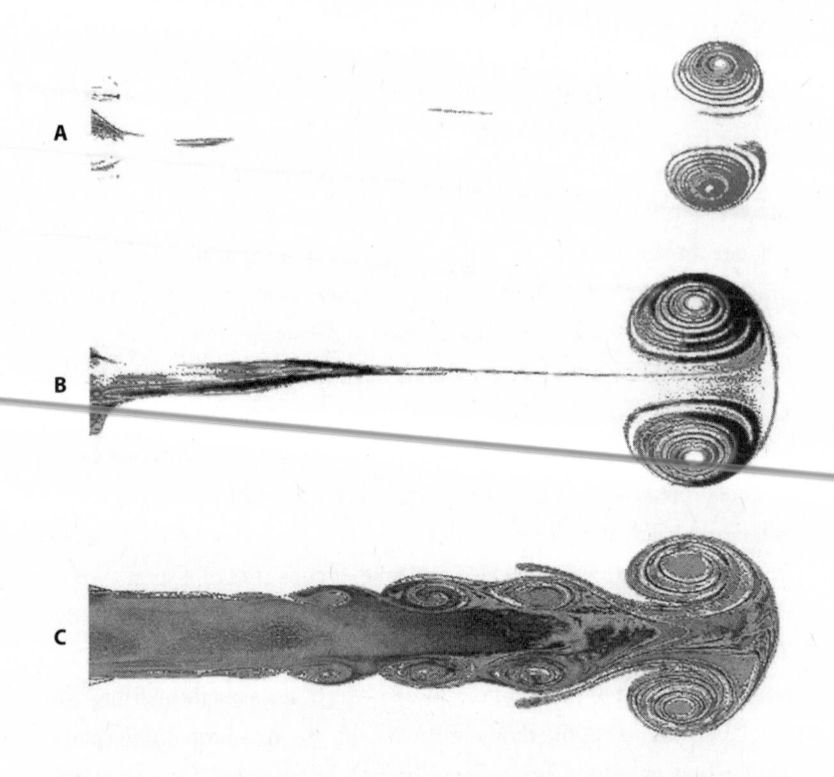

FIGURE 3.3. Dye visualizations of vortex rings created using a piston pushing fluid out of a cylinder. They are ordered in terms of increasing cylinder length of water ejected. Vortex B corresponds to the optimal vortex—optimal because it is the largest vortex that contains all of the fluid in a single ring. Vortex A could have contained more energy if the ejection had persisted. Vortex C is leaking kinetic energy into its wake. Courtesy of Mory Gharib.

a soda can has an aspect ratio of around 1.8, most similar to *Sarsia* in Figure 3.2. Within the lineup of jellyfish, John found a distinct difference in wake for jellyfish skinnier than an aspect ratio of 4.0. He called this ratio the "length-to-diameter ratio for optimal propulsion," or the jellyfish ratio. Jellyfish of this ratio generated clean wakes as in Fig. 3.3B. If a jellyfish was more than four times taller than it was wide, its wake was messy, and the jellyfish would be better off being shorter and putting less fluid into its jet (Fig. 3.3C). Similarly, if the jellyfish was shorter than the jellyfish ratio (Fig. 3.3A), it was acting suboptimal in its stroke: if it were taller, it could push more fluid into the jet and travel farther. Jellyfish that were in between the dinner plates and bullets—call them optimal jellyfish—were traveling the farthest distance considering the energy put into each stroke. The jets that optimal jellyfish produced (Fig. 3.3B) packed the most energy possible into their wakes, no more, no less.

In the years preceding John's jellyfish study, the jellyfish ratio appeared over and over in biology, from heart valves to squid. In fact, the ratio had originally been discovered outside of a biological context. In 1998, John's PhD advisor, Caltech professor Mory Gharib, had discovered the idea of optimal creation of vortices in experiments with pistons that would push fluid out of cylinders. Backed by evidence from the jellyfish experiments as well as from Mory's classical experiments, John was confident that this jellyfish ratio was precisely the number he was looking for to design into his submarine.

When John returned to Caltech, he and two graduate students, Lydia Ruiz and Robert Whittlesey, worked on their submarine. Their primary task was setting the timing of the submarine's inlets, holes on the side that could be opened and closed to allow fluid to enter the submarine's central cavity. The fluid in the central cavity would then be ejected with a propeller. They found the submarine's energy expenditure was a strong function of the length of the pulses it pushed out behind it. If they made the inlets open frequently, the pulses were short and fat, like a pancake. By holding the inlets open for long periods of time, conversely, John and his team could generate long skinny jets, like a hot dog. An intermediate timing turned out to be the most effective. Squirting fluid jets with an aspect ratio of 4:1 released just the right amount of fluid to maximize the distance traveled for the energy used. A longer jet would achieve diminishing returns because fluid would be pushed out in a messy wake instead of a clean vortex. The submarine pushed itself in spurts, like the optimal jellyfish, leaving clean vortex rings behind it. Such wakes packed the most momentum possible without becoming messy and noisy—features that submarine designers want to avoid to deter detection. John showed that the jellyfish ratio could be incorporated in the design of things that didn't look at all like jellyfish.

Consideration of shape is important in getting water to be ejected optimally. As we will see in our third example, it can also affect how well a snake glides through the air.

* * *

It was 1997, and University of Chicago biology graduate student Jake Socha shook hands with the police officer, who handed him a canvas bag tied off with string. The contents of this bag was exactly why Jake had traveled halfway across the world, from Chicago to Singapore. Over the next few years, he would come to southeast Asia regularly for transactions like this. He was in pursuit of the five members of the flying snake family, the most adept of which was the green and yellow mottled *Chrysopelea paradisi,* the paradise tree snake. This time was his first trip to Singapore, and he had come almost entirely unprepared, with only a vague plan to meet people who would help him find and film these snakes, in order to understand how a snake could fly. Jake's first task was to find the snakes, which were particularly elusive, especially in their home, the Singapore rain forests. They weave their way through the canopy, seeming to just disappear. Luckily for Jake, the Singapore Zoo had a long-standing relationship with the city's police, who brought the zoo flying snakes on a regular weekly basis, as if they were pizza delivery.

For years, a few elite members of the Singapore police force had been trained not only to catch criminals, but also to subdue snakes. Singapore was surrounded on all sides by rain forest, and it was inevitable that snakes, from boas to flying snakes, would slither into the city searching for food. Jake had seen it himself on a tour of Singapore when he found *Chrysopelea* on the balcony of a tall tower. As he approached the snake, it simply took off, its shimmering green body simply disappearing. The snakes had evolved to climb trees, so they had no problem scaling the trellis of tall structures, in search of food. One of their favorite foods, the gecko, could be found on the ceilings and walls of restaurants and apartments. Geckos were easy prey, their forest colors sticking out easily on beige walls. After a meal, the flying snakes were known to take the quick way down from a building. Once, a jogger nearly missed a flying snake passing by his head before it landed in the middle of a busy intersection, stopping traffic. In such cases, the first to be called were the Singapore police, then the zoo; and then the zoo would call Jake.

Jake had first learned about flying snakes when he was applying for graduate school. When he was interviewing at the University of Texas at Austin, he met a biology graduate student, Jimmy McGuire, and his advisor, Robert Dudley, who had long been interested in flight. Jimmy's laboratory contained a series of cages, warmed by heat lamps. Inside one was *Draco*, an ordinary-looking green lizard, about the length of his hand, with large yellow eyes. They can do this neat trick, Jimmy said. He opened the cage and gently pried the lizard's fingers from the bars, cradling him in one hand. Watch this, he said. With the other hand, he gripped the lizard by its tail and then let his hand free-fall through the air. The lizard immediately flared open its wings, flaps on the sides of its body consisting of ribs with webbing stretched between them, like duck's feet. This was Jake's introduction to flying reptiles. From then on, Jake was hooked. How could reptiles fly?

The rain forests of southeast Asia are lush, and life centers around the trees. The trees grow tall and full, blocking out the sun from the undergrowth. The network of branches between trees is maze-like, requiring dexterity from the forests' numerous inhabitants. Birds and winged insects fly between the branches. Apes brachiate, swinging from branch to branch with a motion like a pendulum. Other animals leap, and among them, a few augment their leaping with gliding. Body surfaces are appropriated as wings, often in imaginative ways. Flying tree frogs have particularly large feet; webbing expands their feet even further when they glide. Flying squirrels glide using their loose skin, resembling a long trench coat. To extend their leaps bit by bit, their bodies evolved more and more of this loose skin to better deflect air and control their fall.

In typical locomotion, such as walking, energy must be expended continuously to fuel movement. In comparison, gliding is nature's roller coaster: energy is expended only at the start, when the roller coaster is pulled up to its apex. Flying snakes climb trees up to 50 meters tall, comparable to the height of the world's tallest wooden roller coaster, the Colossos in Heide Park, Germany. An average gliding snake has a gliding ratio of 2, which means that for every one meter the snake falls downward, it sails forward two meters. Thus a snake that climbs a 50-meter tree can glide to a point 100 meters from the base of the tree, in just several seconds. This is much farther and faster than the snake could go if it tried to slither, especially since the forest floor is littered with obstacles and, even worse, predators. The snake's gliding ratio is 10 percent larger than the value for flying squirrels and nearly twice as good as the flying frogs. Although the flying snake has a high gliding ratio, it seems to break the rules set by other gliders. Specifically, it lacks obvious gliding surfaces, besides its belly, which seems too narrow to redirect the flow of air. Although air is lightweight, a large enough gliding surface and a high enough travel speed can generate enough aerodynamic force to balance one's weight. Balancing weight means that the force of gravity is offset, and one falls more slowly. This is the basic principle of gliding. The glider must have large surfaces to deflect the air. Strangely, the flying snake has no obvious gliding surfaces. Its body cross section is round. The flying snake would seem as good a glider as a stick.

How does a stick take off? How does it land? For an airplane pilot, these skills require years of training, high visual acuity and coordination, and of course the right landing equipment. In nature, landing equipment usually means legs. Gliders such as the frog and the squirrel are also good leapers, a skill that they use to clear the tree from which they take off. Like a cliff diver, they need a leap to begin the forward momentum that pushes them past the branches and foliage immediately below. Flying snakes do not have legs, so it's not clear how they can leap to begin their glide.

Another important part of biological gliding is that the animal must land gracefully, which is to say, safely, securely, and purposefully. In a forest, the glider is most likely going to make contact with a tree or some other object. A good grip helps. The tree frog has sticky toe pads to grip tree leaves, a squirrel has claws to grip tree trunks. Strangely, the flying snake dispenses with claws, pads, and any appendages. With no wings and no clear landing gear, the flying snake seems like the worst possible glider.

To understand how the gliding snake rose to these challenges, Jake had to overcome his fear of heights. Near the entrance to the zoo was a large clearing, about the size of two tennis courts. Jake hired a Singapore street scaffolding service to build a 10-meter-tall, three-story tower out of hollow metal beams. It was going to be a drop tower for snakes, based on similar experiments done by a scientist over a generation ago.

In 1970, Ron Heyer performed the first documented drop tests with snakes. He climbed a 41-meter-tall, 11-story tower and dropped both non-gliding and flying snakes from the top level. If a person were to fall from 11 stories, the chances of death would be over 95 percent, the remaining 5 percent survival occurring if the person hit a bush or some soft object on the way down. Unspecialized non-gliding snakes have a better chance of surviving a fall from this height than humans because of the scaling of terminal velocity with body size. As said by British biologist J.B.S. Haldane, "You can drop a mouse down a thousand-yard mine shaft; and, on arriving at the bottom, it gets a slight shock and walks away, provided that the ground is fairly soft. A rat is killed, a man is broken, a horse splashes."

Ron found that non-gliding snakes simply fell straight down, landing at most 12 meters from the base of the tower due to wind carry that arose at those heights. Surprisingly, one of the two flying snakes studied had a spectacular drop: after being released from Ron's hand, it flew up to 30 meters from the base, more than three times farther than its non-gliding counterparts, suggesting a high gliding ability. Ron also noted that the performance of the two flying snakes tested was highly variable: in one instance, one of them did poorly, landing within five meters of the tower. He attributed it to adverse wind conditions, cold weather, and the snake's relatively large size, more than 200 grams. This may not seem very heavy, but it is for an object that has to glide through the air. In comparison, the flying snakes that Jake studied were 2 feet in length, and only 30 grams.

Jake hypothesized that the poor performance of one of Ron's flying snakes had to do with the circumstances of the takeoff. The importance of takeoff is also well known to pilots in the hang-gliding community. Most hang gliders are pulled aloft by an airplane, which releases them high up and away from foliage, mountains, and other obstacles. A riskier way to take off is to take a running start from the edge of a cliff. Such pilots must make sure the wind is blowing in the right direction and that they jump far enough off the cliff. Even then, wind conditions can change suddenly, making jumping off a high place to glide always a bit of a gamble. Jake speculated that Ron's flying snakes suffered from poor takeoffs because Ron dropped them by hand. In Jake's experience, flying snakes use gliding as a last resort if they need to escape. He had to be careful to let them initiate their own gliding, rather than pushing them off the tower, which, as in Ron's experiment, could lead to poor flights, and even death. Although Ron did not specify whether animals died in his experiments, it's quite possible that the 41-meter drop led to injury.

Once Jake had figured out how to get the snakes to initiate gliding, he had to plan how he would collect data from their flight. During this time, he took walks through the rain forest. He imagined a flying snake gliding from tree to tree, far up in the canopy. It could travel over 20 meters in a matter of two seconds, meaning that it was traveling over 9 m/s or 30 mph. It was fast. It was also hard to see. The snake was thin, green, and matched its backgrounds quite well, making it almost impossible to follow with a camera. Moreover, the flexible body of the snake allowed it to contort itself dynamically into a number of configurations as it flew through the air. Imagine trying to follow the shape of a kite's fluttering tail as it is blows across the sky.

To understand how the snake glided so well, Jake needed a three-dimensional view of the snake. To accomplish this, Jake recruited Tony O'Dempsey, a snake enthusiast from Singapore whose day job involved making three-dimensional maps using airplane-mounted cameras. Tony convinced Jake that a single camera would be insufficient. Cameras only view their subject from a single direction, which can be limiting for objects of different shapes—or, in the case of flying snakes, that in effect change shape. For example, from the front, both squares and cubes look the same. To distinguish these objects, stereo vision is needed, as is permitted by our eyes. Spaced slightly apart, each eye sees an object from a slightly different perspective. Our brain combines the images together to determine the object's shape, orientation, and position in space. Even our eyes are limited, however, because we cannot resolve the shape of an object in the shadow zones, such as, say, the back or underside of the object. To obtain these views, even more eyes are needed at different positions. To resolve the shape of the snake, Jake and Tony built a 3D filming setup composed of a pair of high-speed cameras, spaced apart for stereo vision, just like human eyes. (On later trips, he would upgrade these cameras to a ring of four to six cameras to decrease shadow zones.) The cameras were set up at the top level of the tower from which the snakes would be launched. At the bottom of the tower were three local biology students from the University of Singapore, Tom Chong, Wendy Toh, and Norman Lim, who would be the catchers for the flying snakes, making sure to pick them up before the snakes slithered away.

Once the cameras and the snake-catchers were set, it was time for Jake to climb the tower. He put his brown bushy hair into a ponytail and cleaned his circle-framed glasses. He tucked a cotton sack with a snake into his belt and began climbing the tower. The tower had no ladder. Instead, a thin column of rungs, just a hand's-width wide, were the only sure handholds. He grabbed the rungs with one hand, and with the other, he grasped the bars making up the scaffolding platform. One by one, he hauled himself up all 30 feet of the scaffolding. He could feel the tower sway back and forth under his weight. The heights made Jake uncomfortable, and to top it off, *Chrysopelea,* like all snakes, were biters. They had two sets of small concentric teeth and from his experience handling them, they left small pinpricks of blood if disturbed. As he climbed, he tried not to jostle the snake too much.

At the top of the tower, Jake gently opened the cotton sack and removed the flying snake. He made sure to grab it closer to its tail to avoid frightening it. On a previous trip up the tower he had brought a long branch, and this time he extended it and tried to nudge the snake to the far end of the branch. The snake lashed out at his arm, giving him a little nip. Eventually, Jake managed to convince the snake to keep on going. The snake looped its body around the branch and hung down, its body in a J-shape. This particular snake lingered there for quite a while, turning its head like a periscope to observe its surroundings.

Gliders are more limited than flapping fliers such as birds. Birds can change their height with the input of energy provided by flapping their wings. Gliders, on the other hand, start with all the energy they are going to get. They can only land in places that are lower than the initial takeoff point, and within the limits of the gliding distance. Like an airplane dropping a bomb, the flying snake must scan its surroundings for possible places to land. Unlike a bomb, the snake can control its direction a bit as it glides, and it must also look for obstacles that it

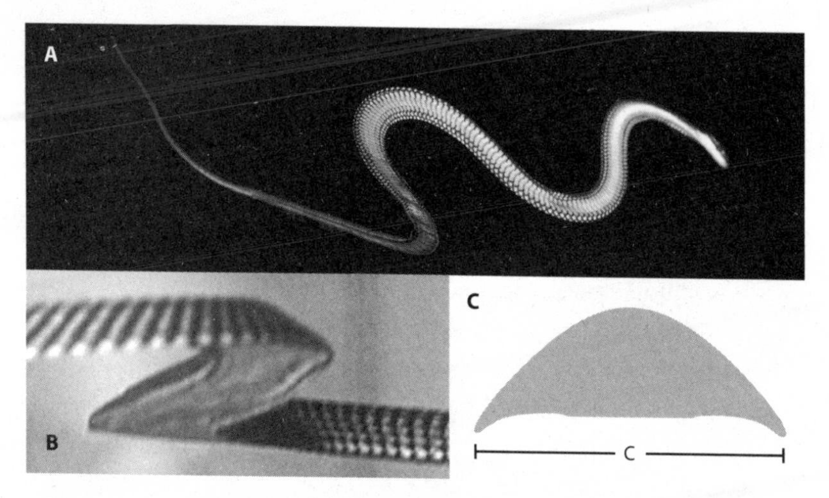

FIGURE 3.4. The flying snake, genus *Chrysopelea*, leaps from trees to glide up to 200 meters through the forest canopy. The snake's ribs are hinged, and when it expands them, it assumes a triangular cross section that reduces the effects of aerodynamic stall. Courtesy of Jake Socha.

will have to maneuver around. Lastly, the snake must be sensitive to changing wind conditions. A sudden gust might throw it back onto the surface from which it leaped. Experienced hang gliders know to avoid taking off in unpredictable wind conditions. Takeoff takes experience and guts.

The snake finally took the leap (Fig. 3.4A). Like a rubber band shot from a finger, it accelerated up and away from the branch. It first straightened its body, transforming into a spear. Then the snake flattened like a cobra's hood. Its ribs, initially pointing toward the ground, swung outward laterally like wings. Its width doubled, giving the snake a gentle concave shape. The snake had transformed its entire body into a wing. The snake dived with its head angled toward the ground. The world, initially still, began accelerating at a terrifying 10 meters per second every second. It's so fast that one second of acceleration results in an increase of speed from zero to 18 miles per hour. If we were in free fall, we would likely feel disoriented as the water in our vestibular canals began floating.

This free-fall acceleration may be frightening to the snake the way it would be to us, but it's necessary. The snake's ribs could only help generate lift if the snake ramped up its body speed. The free fall lasted for a distance of more than two meters as the snake gained speed. As it sped up, it readjusted the orientation of its body in space. Originally angled downward, it lifted its head and lowered its tail, to make its body become more horizontal. It configured its body into an S-shape and began undulating as if swimming through the air. This was a specialized gait for the air only, with a lower frequency and higher amplitude than its motion on the ground.

Now with its ribs outstretched, and in its characteristic body motion, the snake began to change from a stick to a true glider, traveling with its body around 30 degrees relative to the horizontal. The snake's trajectory shallowed as it started generating more lift. If you had been standing below the snake, it would have appeared as if it were going fall on your head, then it would seem to pull up and pass over you. The snake began to travel forward more and downward less. Eventually, the snake reached a nearly steady traveling speed and direction.

For every 15 meters of fall, the snake could travel forward 30 meters. Although it was flying along at nearly 20 miles per hour, it was still alert to its surroundings. Jake's tower was in an open field without obstacles. But in the wild, the gliding snake likely encounters obstacles such as tree trunks in its path. Typical fliers turn by banking their bodies. As Jake observed, flying snakes have their own version of the turn. The snake, which undulated its head back and forth, waited until its head was pointed in its desired direction. Then it simply turned the rest of its body with it, like steering a car.

Eventually the snake came so close to the ground that it was forced to land. However, unlike a car, it could not simply turn on the brakes. Instead it oriented its body to brace for the landing. Then it landed with its tail first and head last. In later experiments, Jake would also observe flying snakes landing on a branch, where the snake has a different method. There, it aims its body at a branch around which it can wrap its tail like a whip. This wrapping motion allows it to slow down as gradually as possible, reducing the landing forces on its body.

From what Jake could see in his videos, the cross section of the snake was no longer circular once it started gliding. It was triangular, with beveled edges and a slightly concave bottom, shown in cross section in Figure 3.4B–C. A wing in cross section is called an airfoil, and this was the first time that airfoils were observed in a snake.

The shape of airfoils has an important role in the history of flight. Although the history of aviation has spanned over two thousand years, the foundation of modern aerodynamics was based on the work of English engineer George Cayley. In the Science Museum in London is a silver disk on which, in 1799, Cayley engraved a sketch for a fixed-wing aircraft, which would become the basis for successful aircraft in the next hundred years. On the reverse side of the disk, he also drew the aerodynamic forces on the wing, separating the forces into weight, lift, drag, and thrust. In the ensuing years, Cayley would build, design and test a number of gliders with fixed wings. The first wings for such gliders were simply wooden boards. He built a "whirling arm apparatus" to whirl the boards about at high speeds, speeds high enough so that he could measure the resultant forces on such wings.

The whirling arm experiments could demonstrate the principle forces that act on an airplane wing. You can recreate these forces by sticking your hand out of the window of a moving car. The weight of your hand creates a gravitational force pointing down. This force is what the airplane must overcome if it is to fly. The airplane generates thrust, which pushes the plane forward. Thrust can be achieved using propellers or jet engines in airplanes and flapping wings in birds, bats, and insects. For your hand, thrust is achieved by the motor of the car pushing your hand through the air. As air rushes past your hand, two aerodynamic forces are generated. The first force is drag, which makes your hand want to fly backward. The second force is lift, which, at sufficiently high traveling speeds, keeps the airplane in the air.

Of course, not all objects that are thrown through the air generate lift. No matter how fast a pig runs, its body cross section will never generate enough lift for it to fly. The most important aspect of lift generation is the wing's orientation. This discovery was made when Cayley put flat wooden boards, acting as wings, in wind tunnels. If the wooden board is held horizontally, the airflow splits evenly around it, flowing in equal portions around the top and the bottom. Due to symmetry, no lift is generated. However, once the leading edge of the board is inclined upward, the air begins to change direction. At low air speeds, no turbulence is generated and air neatly follows the top and bottom edges of the board. In the wake of the board, the air flows backward and downward. This redirection of the air has two consequences: the board is pushed both backward and upward. These directions are associated with generating drag and lift, respectively. Thus, flight is quite simple: it is simply a matter of tilting the wings. The amount of tilt is called the angle of attack, and it can vary between zero and 90 degrees. If an airplane wants to achieve more lift, it increases its angle of attack.

Fundamentally, airplanes fly by directing incoming air downward. This redirection of momentum creates a lift force on the wing. This principle works great up to a large angle of attack of the wing. As the wing increases in angle, fluid no longer smoothly follows the top and bottom of the wing, but separates. The wake becomes turbulent, a state where fluid does not follow straight or curved trajectories, but follows crisscross trajectories, characterized by eddies. Without fluid flowing smoothly down the wing, the wing no longer has much lift. It stalls.

If lift is the angel of flight, then stall is its demon. If a pilot wishes to climb, she must increase her wing's angle of attack to increase lift. However, by increasing past 15 to 20 degrees, lift forces abruptly cease as the wake becomes turbulent and the wing stalls. When Jake became an assistant professor at Virginia Tech in 2008, he worked with Pavlos Vlachos and others to conduct water tunnel tests of the snake's airfoil shape. The snake's airfoil was able to delay stall up to an angle of attack of nearly double what airplanes can take, at nearly 25 degrees. Moreover, its behavior in that region was robust. Increasing the angle further did not cause catastrophic stall, but a steady maintenance of lift forces.

Future studies are needed to further understand the benefits of a triangular cross section like the flying snake. Such cross sections are unlikely to be used in fixed-wing aircraft, whose airfoil shapes have been honed for years. Instead, the snake seems to gain advantage because its body is an S-shape. As air flows over its head, vortices are shed that may interact with the downstream parts of its body. The snake might react to these air flows by oscillating its body to gain further lift. This area of research is called *fluid-structure interaction*, and its consideration is relevant when designing rows of identical structures such as light posts or telephone poles.

* * *

In this chapter, we learned about animals that can use specially shaped body parts to influence the flow of fluids in ways that benefit them. Shape is not the only way to influence flow, however. By evolving specialized surface features, such as hairs or scales, animals can also influence the way fluids flow around them. In this next chapter, we will see how scientists study such small features.

CHAPTER 4

Of Eyelashes and Sharkskin

Since 1972, allergy doctors had relied on a list of symptoms to determine if a child had an allergic disease. These symptoms included dark rings under the eyes due to poor sleep, creases on the nose due to frequent rubbing, and "long silky eyelashes." This latter feature had no explanation, although along with the other symptoms, it had been used in medical practice for years. In 2004, Israeli allergy doctors decided to put it to the test. They measured the eyelash length of patients that were having allergic reactions to house mites, and those that had no allergic symptoms. They found that children with allergies had eyelashes 10 percent longer than if they were allergen-free. It was an intriguing discovery, and the authors had an idea about how it happened. They speculated that specialized mast cells at the edges of the eyelid acted as "central switchboards," releasing chemicals that cause hair growth in response. When I read this study, I agreed with the authors that what appeared as correlation was probably causation: patients exposed to house mites would grow longer lashes in response. However, I also felt there was a missing link in their explanation. Specifically, what benefit was conveyed by having longer lashes? Other studies also suggested that eyelashes play a role in maintaining human health. Madarosis, a condition associated with a lack of eyelashes, was also correlated with higher rates of eye infection. How could our eyelashes be protecting us?

Animals are surrounded by clouds of dust. The faster the animal moves, the faster the dust deposits, just as a car on a highway is pelted by raindrops. Dust by itself appears harmless, but it can carry bacteria and viruses. Particularly large particles can also cause damage to the sensitive eyes of mammals, which are wet and vulnerable. How do animals protect their eyes as they move about? These were all difficult questions, so I started with an easier question. Which animals have eyelashes and how long are they?

To determine which animal had the longest lashes, I recruited Peter Mercutio, a student from my fluid mechanics class and a Brooklyn native, to take a trip to the American Museum of Natural History. He was not headed for the area open to the public, but to the zone set aside for research, filled with priceless collections, heavily guarded by layers of security. To enter this area, we filled out a lengthy application form, as if planning to visit a foreign country. We detailed our purpose, the names of every species to be examined. It was made clear that Peter would have an escort at all times, and that he could only see a small part of the collection. Once inside, however, it's easy to forgot all about the effort required to get there.

One of the pleasures of being a scientist in animal locomotion is seeing the backstage of a natural history museum. Hidden entrances and secret elevator buttons take you many floors below ground, to areas inaccessible to the public. In a labyrinth of earthquake-proof rooms are shelves and cabinets filled with skeletons, pelts, and feathers collected by America's first pioneers, explorers, and even ex-presidents. Now, the collections continue to grow through the work of curators and scientists who travel around the world in efforts to capture the diversity that nature has to offer before it disappears.

Peter spent a week in the catacombs of the museum, carefully examining animal pelts for eyelashes. (Plate 7 shows photographs of goat eyelashes later taken at the Atlanta Zoo.) Peter returned with photographs of the holes where the eyes had been and measurements of the eye hole sizes and the eyelash lengths. If the eyes were circular, we reported their diameter; when elliptical, we reported the average of the eye height and width. The hedgehog had an eye only 0.5 cm wide, the width of the eraser on a pencil. The giraffe had the longest lashes, almost 1.5 cm long, with an eye nearly 4 cm wide, about as wide as a cappuccino cup. The human eye is only 2.4 cm wide, and generally has a single layer of eyelashes, unless you are the actress Elizabeth Taylor, who had a genetic mutation causing her to have double layers of eyelashes.

When we considered the data all together, we were surprised to find that the length of lashes could be easily described mathematically. On average, the lashes had a length of one third the width of the eye. We call this a constant proportion, or isometry, as we observed in the urethras in our last chapter. In other words, a giraffe's eye is just a larger version of a hedgehog's eye. As the eye gets bigger, so do the lashes, in proportion.

This was a very strange finding indeed. We had expected the eyelash lengths to be all over the place. The density of eyelashes, ranging between 20 and 80 hairs per centimeter, showed no such trends. But somehow, the range of sizes of animals, their habitats, and their widely varying lineages led them all to the same solution of having eyelashes of the same length. What was this optimal eyelash length, and what benefit did it provide? This was an open-ended question, and I did not have answers, only hunches.

When I came to Georgia Tech in 2008, I became close friends with Alex Alexeev, a tall thin Russian who told depressing jokes and listened to Ravel's *Bolero* for hours on repeat. He was educated in Moscow and Israel and had become an expert on brute-force numerical simulations, those that calculate the motion of millions of tiny fluid particles throughout a flow field. Alex's techniques were useful for calculating motion in three dimensions, which is difficult to do with pen and paper. No problem was too big for him, and we started collaborating. Our first paper together was published a year before I started thinking about eyelashes. It was a study of the growth of the world's largest pumpkin, which was around 2000 pounds. At that size, the pumpkin's own weight forces could stretch its tissue, enhancing growth in certain areas, resulting its being wide but not tall. Alex did these computations in his computer cluster, a set of 50 computers in a hot little room just at the edge of campus. An air-conditioning system was needed to prevent the computers themselves from melting.

I told Alex about the constant eyelash proportion I had observed. Alex wrote a code to predict the resulting flows from wind directly impacting an animal's eye. This is similar to the flow that occurs when the animal walks forward. Once the code was written, he could easily lengthen the eyelash to observe its effect on the flow around the eye.

Alex's computers churned day and night for an entire week. At the end, he found that lashes of the 1:3 proportion acted like a windbreak for the eye, reducing flow at the eye surface. Alex's work shows that computers are a critical part of understanding biological systems, especially when we have no clue where or how to start doing experiments. The effect seemed to be real, but no one would believe us unless we also had experimental evidence. We decided to build an eyelash wind tunnel.

Finding a student in mechanical engineering to investigate the aerodynamics of eyelashes was not going to be easy. But in every entering class of fifty students, I noticed that there was one student that had inclinations toward biology. When I first met Guillermo Amador, he had long hair, and wore leather wristbands and a faded Venezuelan soccer T-shirt. He had seen my snake work in *National Geographic* and he had come to find out more about working with animals. He spoke thoughtfully, and his eyes looked wise behind his glasses. He agreed to take on the study of eyelashes for his doctoral thesis.

Inspired by Alex's simulations, Guillermo began building a wind tunnel to study eyelashes. He wanted to generate winds associated with a leisurely walk, about one meter per second or 2 mph. This was the speed that we approximated animals would walk: as they walked forward, they generated a slight breeze impacting their eyes, at least for animals with forward-facing eyes. The speed was slower than any wind tunnels on the market. The professional wind tunnels to study aircraft models could blow air at up to 400 mph, and could not stably generate the speeds that we needed for our eyelash testing, so Guillermo built his own wind tunnel. An old desktop computer fan sucked the air through the tunnel and an acrylic tube lined with drinking straws smoothed out any eddies. The whole device could fit on a desktop.

Next, we needed to measure air movement at the surface of the eye. This is where being in a university helps, because they have high-tech versions of everyday objects you have at home. In our bathroom, we step on a scale to measure our weight. A laboratory version of the bathroom scale is called an *analytical balance*, and it can detect the weight of a tenth of an ant. It uses high-precision solenoids, magnets that convert the smallest motion into an electrical signal. Moreover, the weight measured is updated every single second. The sales representative liked to show off the scale with a simple demo. He would place a glass of water on the balance. As it sat there, the numbers on the balance began counting down as the water evaporated.

For years, I had this amazing demonstration in the back of my mind. We usually think of evaporation as unimaginably slow, but here was an experiment that could actually measure it. I realized the analytical balance was the perfect tool for studying evaporation of the eye's tear film. We lined a miniature teacup of water with fake human eyelashes, trimmed to the appropriate length (Fig. 4.1, left). We later tried a mesh screen, which had the same effect as the human lashes and was easier to trim (Fig. 4.1, middle). We stacked the wind tunnel on top, blowing air downward, and faced the cup so that it would feel

FIGURE 4.1. Eyelashes prevent the eyes from drying. To measure the effect of variable-length lashes, small cups of water are protected by eyelashes of various types, including human-hair eyelashes (left), a synthetic mesh wall (middle), and impermeable paperboard (right). A wind tunnel pushes air downward onto the eyelashes, and the weight of the water is measured using an analytical balance.

the incoming wind. We turned the wind tunnel on and waited. Even with the wind tunnel on, it was a slow process to evaporate the water in the cup: it evaporated at a rate of 1 mg per minute, or the weight of one ant per minute.

We were shocked by the results. The eyelashes had a dramatic effect on the water evaporation. A cup with no lashes took ten minutes to evaporate. By adding lashes of the optimal length, we could lengthen this duration to twenty minutes. Strangely enough, these effects disappeared once we used longer lashes. Long lashes had deleterious effects, as bad as if there were no eyelashes at all. We surmised that animals with eyelashes of the optimal proportion would experience half as much evaporation of their tear film as animals without lashes. With half of the evaporation, they would need to blink half as often. Humans blink about every six seconds, so in a lifetime, eyelashes could save you blinking over 100 million times.

This optimal eyelash length was only relevant if the lashes were porous. We did experiment with replacing the lashes with a miniature toilet paper tube (Fig. 4.1, right). The tube reduced the evaporation even more than lashes of the same length. However, no optimum eyelash existed: instead, the longer the tube, the less evaporation there was. The problem with the tube is that it did not allow light to penetrate the eye, and would be difficult to clean—just some of the reasons why animals did not evolve with toilet paper tubes around their eyes. We thus returned to studying porous lashes.

The lashes shielding the eye could also potentially reduce the deposition of incoming dust. We studied this ability by seeding the wind tunnel with humidifier mist, which consists of small drops of water nearly ten times smaller than a human hair width. We doped the drops with green fluorescent dye to better measure their deposition. We found that the optimal eyelash length reduced deposition of these drops by a factor of two. This shielding ability against incoming particles could be particularly important in dry areas such as the desert where airborne particles arise in greater numbers. In fact, desert animals have other adaptations besides eyelashes of the right length. Camels have multiple layers of eyelashes that further block airflow; and when their eyelashes are insufficient to block particles, they close their nictitating membranes, or transparent eyelids, which allow them to walk through dust storms.

Alex's simulations could output the *flow streamlines*, the particular trajectories that air follows when it is directed toward the eye. By increasing the length of the lashes, we could see the effect on the streamlines. Without lashes, dry air immediately impacts the eye, stealing water molecules from the eye surface. This air also brings with it dust that can contaminate the eye. The shortest lashes act as a speed bump for the air. Incoming air veers away from the eye earlier, and a stagnant layer surrounds the eye surface. However, long lashes begin to act as a funnel. Fast-moving air far from the eye is entrained into the funnel and the eye experiences more air flow. This is because air flows more easily along the lashes than between them. The compromise between the speed bump and the funnel is what creates the optimal eyelash length, which maximizes the shielding effect of the lashes.

Eyelashes are an example of an air filter that is low-effectiveness but also low-powered and maintenance-free. Human-made filtration systems are much higher in effectiveness—they remove many more particles, but at high energy and maintenance costs. Eyelashes don't use any energy because they take advantage of the air flows generated from an animal's forward motion. Your home's air-conditioning requires a pump to push air through a paper-based mesh filter. Because the mesh is so thick, a substantial amount of energy is used to resist the drag of air pushing through the small holes in the mesh. Over time, this constant ramming of air clogs the filter with particles and requires it to be replaced regularly. In comparison, eyelashes do not need to be replaced because they are not absorbing particles. Instead they simply deflect air flows around a target of interest. By redirecting the flow of air, they can reduce the animal's energy in blinking and the impaction of particles.

In general, we are not good at building systems that work in the low-effectiveness but low-maintenance regime. Currently, solar panels around the world lose 6 percent of their energy collection every year due to the accumulation of dust. Moreover, these energy losses are worse around the equator, where solar panels are envisioned to be most profitable to install. At the moment, the only solution is cleaning each panel by hand with a squeegee. Like many human systems, a squeegee works with nearly 100 percent efficiency to remove dust. However, it requires each solar panel to be maintained by hand, a method that may be difficult to scale up. Eyelashes and other protective structures may be one solution. They would allow light to reach the panels, yet reduce the number of times they need to be cleaned by half.

Eyelashes are small structures that protrude into the air flow around an object. They are one of an infinite number of such devices in nature. Examples are everywhere, from the scales of snakes to the hairs on water striders to the feathers on birds. Because living things are made of individual cells, they have the potential to evolve very fine structures.

What use are such fine structures? Can small bumps, like those on a golf ball, really influence the path of an object through a fluid? The answer is yes, but it took nearly a hundred years to get to this answer.

* * *

In 1744, The Royal Academy of Sciences of Berlin began holding essay contests on a variety of topics; the topic in 1749 was the flow of fluids. In response to the fluids contest, a French mathematician Jean le Rond d'Alembert used the accepted principles at the time to write a remarkable mathematical proof: he proved that drag on a submerged body should be zero. This idea, which became known as d'Alembert's Paradox, was the culmination of years of theoretical work on ideal fluids. The paradox tore the fluid mechanics community into two factions that are still separate, even today. One faction was the field of hydraulics, observing phenomena that could not be explained; the other was theoretical fluid mechanics, explaining phenomena that could not be observed—as much later described by chemistry Nobel laureate Cyril Hinshelwood. Each of these fields plodded along in its respective direction. The former took measurements based in empiricism, while the latter proceeded with d'Alembert's calculations in efforts to resolve this physical impossibility.

In 1904, over 150 years after d'Alembert's Paradox was proposed, the controversy was finally resolved. German engineer Ludwig Prandtl proposed a missing link in d'Alembert's essay, the viscosity of a fluid, a property denoting the "thickness" of the fluid, its resistance to flow. D'Alembert had neglected viscosity because viscous forces appeared to be a factor many times smaller than the inertial forces of the fluid. However, Prandtl showed that this idea was incorrect. In fact, whether viscosity is important or not depends on how close you are to the moving object.

Consider a moving object in a fluid. The fluid around the object can be discretized into a number of layers, like a deck of cards, which must *shear*, or slide relative to one another, to allow the object to move. The card closest to the object remains stuck to it, an idea called the "law of the wall." The greater the fluid's viscosity, the more difficult it is for each of the cards to shear. As the object moves, one would imagine that viscosity would affect all the air around it, causing all the cards to move just a little bit, depending on their distance from the object. In reality, the air far from the moving projectile is stationary as if it's unaware of the object's presence. It is only a thin layer of air around the object that feels the object at all. Prandtl called this the *boundary layer*. It is through this layer that the object feels most of its drag.

One of the reasons that the boundary layer lay hidden until 1904 was that it is very difficult to observe. When you kick a soccer ball, its motion produces a boundary layer only 0.1 mm thick. It's very difficult to observe, and faster objects have even thinner boundary layers. No matter how thin, it is through this small space that the object interacts with the outside world. Within the boundary layer, the object causes the air to rapidly shear. In turn, the viscosity of the air resists this motion and applies drag to the object. In turn, the object is now influenced by the thin layer of air around it. Physically, this is where small changes in the object's surface can have a tremendous impact on its drag. Theoretically, by adding microscopic bumps on the object, one could influence the boundary layer and in turn influence its drag. While this idea was not rigorously analyzed till years later, at the same time in the world of sports, golfers were beginning to learn about the importance of surface roughness. The idea of using small bumps to reduce drag may have been first discovered in the sport of golf. A golf ball has no propeller, no wings, and no other devices to give it direction or keep it aloft. Instead, it has to rely on the initial drive by the golfer, which lasts for a fraction of a second, to control the flight over the next ten seconds. It is a situation where the ball's interaction with the air is of the utmost importance.

One of the first widely used golf balls was the gutty, invented by Scotsman Robert Paterson in 1848. It was molded like clay from the dried sap of the Malaysian sapodilla tree. The gutty was cheaper than the typical leather-bound balls of its day. The popularity of the gutty was also due to a then unexplained phenomenon observed by experienced golfers. New gutties were smooth, but as they were hit, they developed pockmarks, dimples on the surface of the ball from the repeated strikes of the golf club. As a consequence of the dimples, older gutties attracted attention because they flew farther. In today's golf world, computer-placed dimples can reduce the drag force by a factor of two relative to a smooth ball, meaning that balls can fly substantially farther with the same initial strike.

To understand how the dimples work, first consider a smooth ball. When air hits the ball, it travels around it, hugging its perimeter. The air slows down as it travels along the surface, sheared like a deck of cards by the law of the wall. The air is slowed down so much that by the time it travels all the way around the ball, it is now traveling upstream, in the same direction the ball is going. This generates a vacuum, a low-pressure wake behind the ball, that tends to suck the ball backward, increasing its drag.

A ball with dimples helps to reduce this suction. It does so because air does not flow smoothly around the ball's dimples, but instead mixes in the surrounding air. This surrounding air is moving faster than the air in the ball's boundary layer. The fast air mixes in, like energized partygoers entering a dull party. The slow air is invigorated by the new air and the mixture has more speed to travel around the ball. The wake behind the ball is reduced, and the ball flies with less drag.

There was a good reason why it took so long for dimples to be discovered. They are highly nonintuitive. Smoothing an object's surface is intuitively the best way to reduce drag, be it for objects moving through air or water. This has long been known for boats, which are subject to marine biofouling caused by barnacles and other algae and animals that attach to the bottoms of boats. When such organisms accumulate, they can increase drag by up to 60 percent. More drag is particularly problematic for boats and airplanes that carry all their own fuel. More drag requires boats to carry more fuel. This fuel in turn weighs down the boats, sinks them deeper into the water, and creates even more drag. Thus, more than 40 percent additional fuel is needed to overcome the consequences of marine biofouling. Attached barnacles and other marine animals provide surface features too large to reduce drag. Only very small-scale roughness that remains within the boundary layer can do so.

The idea of using rough surfaces to mix the boundary layer and reduce drag has also been applied to airplanes and cars. The wings of airplanes or the roofs of cars have, instead of dimples like a golf ball, a series of small vanes that act like dimples, mixing in fast air to reduce drag. One of the most commercially successful is riblet tape, so named because its surface resembles rib bones, covered with small triangular grooves each 1/3 the width of a human hair. Riblets have been applied to the fuselages of airplanes, the hulls of yachts, and even the blades of wind turbines. One of the problems of riblets is that it is expensive to coat entire surfaces with them. They must be manually applied, they become degraded over time due to dust wearing the surfaces, and they make it difficult to apply de-icing fluids.

Although clogging and wear has limited the use of such added surface features on commercial vehicles, surface features are a common appearance in nature. How often do you see an animal that is perfectly smooth? On both land and water, surface features on animals can reduce their resistance as they move. A snake's contact with the ground can be boiled down to the frictional behavior of small patches of its body, which are in turn affected by the snake's surface chemistry. Similarly, the drag on a swimming or flying animal in fluids boils down to the interaction between surface features and the passing fluid. This is particularly important in swimming because water is 1000 times denser than air. No animal pushes the envelope more for reducing resistance than the shark.

Sharks play the endurance game because they must swim constantly to keep water flowing through their gills. Moreover, their predatory lifestyle requires them to summon speed when prey is near. The fastest sharks can travel at 20 mph in short bursts, nearly half as fast as the Soviet K-222, the world's fastest submarine. For years, the secret to the shark's speed has been thought to be its skin. In our next story, we will see how to measure the drag-reducing properties of sharkskin.

* * *

At the Boston fish market, Harvard biology professor George Lauder and his two students looked at the one-meter-long mako shark, sitting on a tray of ice cubes. Its body seemed built for speed. A rocket-shaped nose led to triangular fins and a long, angled tail fin, like a hockey stick. George rubbed his hand down the shark's gray, matted flanks. At first touch, he was surprised. The skin felt extremely rough, like the largest-grain sandpaper, although the grains were invisible. The Maori people of Polynesia had recognized this roughness years ago, and used dry sharkskin to sand their boats smooth. The Japanese used it on the handles of their swords to keep them from slipping out of their hands. Under a microscope, the sharkskin had a grain, similar to a cat's fur (Fig. 4.2). Depending on where it was on the body, it has also had varying roughness. The leading edges like the snout and the front of the fins were smoothest. The gills and belly and tail each felt like it had different roughness, as if one were browsing the sandpaper aisle in the hardware store, and seeing the bins of different grade sandpaper. A rough skin seemed a strange material to cover a fish with, especially the fastest fish in the world.

An adult mako shark is about the length of a Volkswagen Beetle, and weighs about 300 pounds. In short bursts, it swims up to 20 miles per hour, and it can leap up to six meters out of the water, the height of a two-story house. George remembered how when he used to go sports fishing, the makos chased their speedboats. The mako does not look

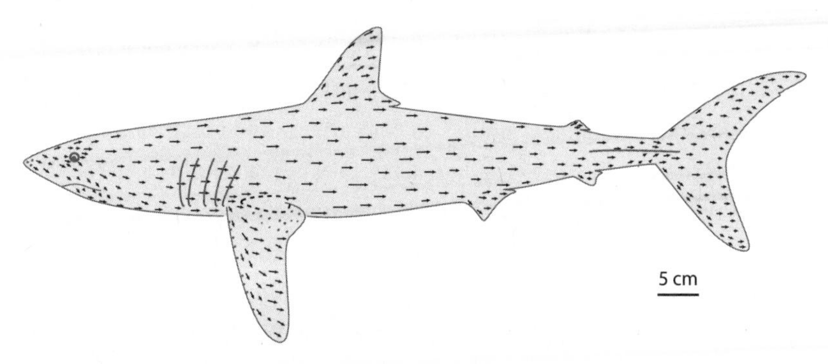

FIG 4.2. Orientation of the scales in the shortfin mako shark *Isurus oxyrinchus*. Arrows indicate fluid flow direction driven by orientation of the scales on the body. Figure modified from original by Wolf-Ernst Reif.

like it's swimming, exactly, but more like it's zipping around like a fly. The shark has *contragility*, the ability to make sharp changes of direction while already in a turn. Submarines and aircraft have high peak speeds, but do not have contragility. The mako easily swerves between billows of seaweed or coral reefs as it chases its prey. If you want to build a device with contragility, you face a fundamental problem. Rapid changes of direction involve moving fins at high angles of attack to the flow. At high angles of attack, fluid does not stick to the surface of the object, but separates from it, generating eddies that increase drag and reduce lift. The mako's swimming ability boils down to its stickiness, its ability to keep the fluid sticking to its sides and moving smoothly as it slices through the water.

For years, sharks and dolphins had captured scientists' attention for their speed. They were thought to have "fast skin," a mysterious surface imbued with drag-reducing properties. Dolphin skin is smooth and shiny, like wet human skin, but much thicker and with a density like silicone. In comparison, sharkskin is matted, and with the texture of sandpaper. Although it is a fish, its scales are microscopic. The first experiments in locating the source of the speed of these animals involved towing dead dolphins and sharks through water behind boats and in water tunnels. Disappointingly, sharks and dolphins were found to have drag coefficients similar to each other, and equal to that of a small dinner plate. This was surprising because sharkskin was rough, while dolphin skin was smooth. Thus, initial experiments with dead sharks did not reveal the scales as having a drag-reducing effect.

Further experimental work was done on building sharkskin mimics. The main player was turbulence researcher D. W. Bechert, who in the 1980s built a series of shark-inspired surfaces at the Institute of Propulsion Technology in Germany. At the time, scientists in the United States had invented riblets based on fluid mechanics reasoning. The Germans hoped to beat the Americans in building drag-reducing surfaces by accurately replicating sharkskin.

Using microscope images of the scales as a model, Bechert hand-sculpted clay models that were 600 times the size of a shark scale, about 6 centimeters in size. He used a pantograph-copy milling machine, an attachment to a regular milling machine that uses a series of linkages to trace the outline of some template and then uses rotary cutters to make a copy that is reduced in size 100 times, to about the size of a pinky nail. Then a plastic casting machine made 800 more copies in polystyrene. The artificial shark scales could be attached to a metal plate simulating the shark's body. Individual leaf springs attached to the plate allowed the scales to rotate, just like scales on a real shark.

The entire setup was tested in a specially constructed water tunnel that flowed oil past the plastic shark scales. It might seem strange to use oil instead of water, but it is a common method in fluid mechanics. The reason is that the artificial shark scales were 600 times larger than the biological shark scale. In order to recreate the same flows and physics, they need to increase the viscosity of the testing fluid to compensate. Imagine if you zoom in really close to a shark scale: at higher levels of magnification, the fluid appears more viscous. That is, the viscous forces depend on the size of the system at play, a concept called *dynamic similarity*. This is the whole reason that wind tunnels and water tunnels provide useful data: scientists can put smaller versions of boats and planes in these systems, and change the speed or viscosity of the fluid in order to replicate real-life situations.

With very stiff springs and the scales aligned, the German researchers found a 3 percent drag reduction. This was disappointing, especially since this drag reduction was far less than achieved by the stiff riblets that the Americans had built. The American riblets consisted simply of a series of triangular grooves, not resembling sharkskin at all. Their drag reduction was 10 percent compared to a flat plate without scales. Bechert and most of the field of fluid mechanics remained mystified by shark scales. Why did the shark have such complex structures when simple triangular grooves would work just as well?

Since then, interest in sharkskin has grown even greater, fueled by the consumer market. In 2000, British designer and former competitive swimmer Fiona Fairhurst turned her attention to making a sharkskin-inspired swimsuit. She designed a series of suits, named the Fastskins and LZR suits. These suits, manufactured by Speedo, covered the body from neck to ankles and were textured with ridges, mimicking the shark scales, envisioned to reduce drag. Speedo claimed the suit could reduce a swimmer's drag by 6 percent. The suits were launched in 2008, in preparation for the summer Olympics in Beijing, where 150 suits were distributed to the world's top athletes. Ultimately, the Speedo wearers broke 23 of 25 world records, and the International Swimming Federation unanimously voted to ban the suits. They argued the suits were a form of technology doping. Like webbed gloves and flippers that improve a swimmer's speed, buoyancy, and endurance, the suits gave their wearers an unfair advantage. The controversy made national headlines, and piqued George's interest in how these suits actually worked. He recruited Johannes Oeffner, a master's student from Germany interested in doing his thesis on sharkskin.

Shortly after he arrived in Cambridge, Johannes began his thesis by driving with another student to a local fish market and purchasing two frozen sharks, a mako and a porbeagle, asking only for the rear halves for ease of transport. He spread a tarp in the back of his small hatchback and loaded in the sharks, which lay in the car like knocked-over bowling pins, and drove them back to George's lab, now housed in a newer addition to the Harvard Museum of Comparative Zoology, a Victorian-style mansion that looked like it belonged to the Addams Family. The sharks were over 50 kilograms total and required a rolling cart to move them from the car to the lab. When they arrived in the lab, the students gathered around to admire the specimens. The flesh was pink, with very little blood, and the skin was very firmly adhered. Johannes sat astride each shark like he was riding it and, using tongs and a razor blade, slowly began cutting the skin off each side. After two hours, he had four pieces of sharkskin, each piece 50 centimeters long and 15 centimeters wide. Although he had tried to cut as close to the skin as he could, there still remained pink shark flesh attached. The skins would have to be cleaned, and now time was becoming an issue because the sharks were steadily decomposing. They had been caught the day before, and he only had so many hours to remove the skin to perform his drag tests.

Johannes used a set of dissection scalpels to peel back more layers of flesh from the skin. Around him, the flesh piled up in pink heaps, like piles of used chewing gum. When the sharkskin was sufficiently clean, a high-speed jet, composed of water and small grains of sand, served to remove the pieces of flesh that were too tightly bound for his scalpel. At the end, he was exhausted but satisfied. He had four rectangles of clean shark leather, gray on one side and white on the other. He brought the pieces to George, who looked at them closely.

George's background studying fish gave him an advantage when deciding what kind of experiments to do with sharkskin. He wanted to make the testing as similar as possible to the movement of a shark. Previous scientists had attached sharkskin to a rigid plate and then measured the drag force as water was driven past the plate. In reality, sharks undulate their bodies and tails to generate thrust. This undulatory motion could give rise to new kinds of flows that could interact with the flexible sharkskin. To generate this motion, George had one of the few apparatuses in the world that could provide a fishlike motion for the sharkskin.

Ten years earlier, George had invited a roboticist named James Tangorra to do a doctoral thesis in his lab. The product of five years of development was an apparatus for generating fishlike gaits. Imagine a motorboat that was not pushed forward by a propeller but instead by a robotic fish tail. In essence, this is what James built. He called it the flapper. It controlled the motion of a single foil, or plate, aligned with the incoming fluid stream, like a flag pointing downwind. The flapper itself sat on nearly frictionless air bearings, as if it were a hockey puck on an air hockey table, floating on air. Consequently, forces on the flapper resulted entirely from its interaction with the fluid. Two motors controlled the motion of the plate in a fishlike manner. The plate could *pitch*, changing its angle with respect to the direction of flow. It could also *heave*, or move laterally, as a fish does to avoid obstacles. George's idea was to give the flapper a new set of clothes, namely one made of sharkskin.

To program the flapper to move similarly to a shark, Johannes and George began with experiments with a live shark. They bought a foot-long spiny dogfish shark from vendors of live fish in the aquarium trade. The fish was trained to cruise inside the *flume*, or water tunnel. A comfortable speed that it could maintain indefinitely was 1.5 body lengths per second, which is just a bit faster than our fastest human swimming speed. To swim at this speed, the shark leisurely swung its tail back and forth about once per second. In the wild, it could maintain this speed for hours on end as it roamed in search of food. Johannes then recorded the dogfish body shape as it swam. To swim at this leisurely speed, it arched its body with a radius of curvature of about a dinner plate. Now that he knew what shape and frequency to program into his robotic flapper, it was time to build the fin.

Johannes planned to use his sharkskin leather to construct flexible flags made of sharkskin. To begin, he needed to find a glue that would work with sharkskin. There were various glues in the lab, including school glue, rubber glue, and hot glue, but Krazy Glue seemed to work the best. To carefully glue the two pieces of sharkskin together, Johannes had to hold the skins clamped between two slabs of wood. After a few near scares of gluing his fingers together, he finally learned to create a nearly seamless edge with skins. Finally, he had gray flags of sharkskin, with the rough sides of the skin facing outward. He attached the flags to the shaft of the robotic flapper, where the underwater foil would ordinarily sit. He then flapped the flapper back and forth, heaving and pitching the flags to mimic the motion of the dogfish shark he had observed earlier. As the sharkskin flag was waved back and forth, the entire flapper was propelled forward in the flume. Eventually, Johannes could tune the water speed so that the flapper held station in the flow. Here the net drag and thrust on the flapper was zero. The water speed at which the flapper held station represented how well the flapper could swim.

Now that Johannes had established a protocol for measuring effectiveness of swimming, he was ready to test the effect of shark scales. To do that, he had to design a *control test* to compare the benefit of the sharkskin to a smooth surface. Figuring out the appropriate scientific control is a subtle open-ended problem that every scientist must face. He could use a flexible plate made of plastic, but tuning it just right to the flexibility of the sharkskin would be difficult. He needed two samples of sharkskin, one with and one without scales. But where would he find a shark without scales?

George suggested that Johannes remove the shark scales from the mako skin. Johannes took the finest grade sandpaper and began sanding the skin in a small plate of water under the microscope. After a few gentle brushes, he had removed the tips of the scales, with only small stubs at the base of the scale remaining. The original scales looked like an array of mushrooms, but after the sandpapering the mushrooms were obliterated, leaving the skin like a battleground. An ultrasonic bath removed any remnants of the sandpaper and scales. Now Johannes was ready to do the appropriate control test.

Johannes attached two sanded sharkskins together like the flag, just as he had earlier with unsanded sharkskin. When he compared the unsanded sharkskin to the sanded one, he found the shark scales provided a 12 percent increase in swimming speed. However, the shark scales only helped if the flag was allowed to undulate. Johannes also conducted tests with both sanded and unsanded sharkskin glued to rigid plates, which prevented the sharkskin from undulating. When the sharkskin was attached to rigid plates, he found the opposite trend. Here, the sanded sharkskin was the better material; in this case, the sharkskin was to the shark as barnacles are to a boat, providing unwanted roughness and increasing drag. It was a mysterious result. It appeared that the shark scales were not useful to the shark unless the shark was undulating. This is an important emerging picture of animal locomotion: the secret to speed and fuel economy emerges from the interaction of body movements and material properties. Such interactions are surprising, and take advantage of nonlinear properties of fluid mechanics.

Johannes conducted tests with synthetic materials attached to rigid plates to compare them to the sharkskin performance. They were all worse than sharkskin. Riblet material provided a 7 percent increase in swimming speed, still only half of the benefit of sharkskin. Speedo swimsuit fabric provided zero increase in swimming speed. Where was the 6 percent increase in swimming speed that the Speedo company had observed on their wearers? This increase, it seemed, had nothing to do with the texture of the surface of the suit, as had been thought. Instead, it likely resulted from the suit's compressing the swimmer's muscles into a more streamlined shape. In any case, it appeared the sharkskin had at least double the abilities of synthetic sharkskin.

Now Johannes had solid data showing that a shark's scales improved its efficiency. The 12 percent increase in swimming speed means that after 10 miles, a shark can swim an extra mile for free. In a typical day a shark may swim a hundred miles because it is constantly cruising in order to aerate its gills. Without the scales, the shark needs to find 12 percent more food to sustain itself.

Where did this 12 percent increase in speed come from? To answer this question, Johannes embarked on flow visualization of the flapping foil, which was his favorite part of working in George's lab, because it reminded him of dance clubs in Germany. He turned off the lights and illuminated the flow around the foil with a laser light sheet. The motion of the fluid was traced out by the movement of small, neutrally buoyant particles called *tracers* that traced out flow patterns like a starry night sky.

As the sharkskin undulated back and forth, it created a vortex near its front edge. The sharkskin managed to draw the vortex in closer to the body, as if it were stickier than smooth skin. The sanded skin also had a vortex, but it was farther away. The center of the vortex was a region of low pressure. This low pressure was so close to the unsanded sharkskin that its low pressure could actually suck the sharkskin forward like a vacuum cleaner. Every time the sharkskin undulated, the vortex was created, and the sharkskin was brought forward. George and Johannes concluded that the scales permitted the shark to stick better to leading-edge vortices, which increased thrust of the foil. To further test the abilities of the scales, George would turn to a new technology known as 3D printing.

3D printing has been praised as bringing about the next industrial revolution. In the field of animal motion, it's also becoming an increasingly important tool for studying highly intricate biological structures.

It involves similar equipment to an ordinary inkjet printer, which uses a motorized syringe to inject spots of ink on paper. A 3D printer rises to different levels above the paper. As the ink dries, the structure rises like a tower of small building blocks. The 3D printer can nearly replicate itself, able to build all its plastic parts, but not yet the electronic components. It can also 3D-print a primitive lathe, which is one of the few other machines that can replicate itself. Just like an inkjet printer can be filled with different colors, a 3D printer can inject materials of different densities, such as plastic, metal, and even liquids. Researchers have printed a fully working liquid-filled piston by printing both plastic and liquid so that the piston does not have to be assembled by hand. The 3D printer can print metal parts for use in airplanes and cars. Highly specialized and expensive 3D printers can print entire carbon-fiber frames for cars. Living stem cells can be ejected like ink to generate artificial tissue such as skin or a heart. Microfluidic channels can flow fluids and carry chemical signals across the stem cells to control differentiation.

The average home 3D printer cannot produce materials that can support large forces. Thus, at this point it can make a good cell phone cover, but not a good chair. The ability of 3D printing to make arbitrarily shaped surfaces, however, is powerful, especially for biology. The power of biological development is that structures can be developed with a high degree of complexity. Pollen, vertebrae, and antlers are all structures that would take days to carve by hand out of wood. However, a 3D printer can make them within minutes, and then hundreds of them at a time.

To demonstrate the ability of the sharkskin, George decided to 3D-print a replica of the skin and test it. George paired up with his postdoc Li Wen. The male shortfin mako shark obtained from the Boston fish market was still in the freezer. They cut a 10-centimeter-square piece of skin using a scalpel and then cleaned it with a water jet. From this square, they cut an even smaller piece, a two-millimeter square. They

FIGURE 4.3. The scales, or denticles, from the surface of the mid-body region in a bonnet head shark (*Sphyrna tiburo*), as shown with scanning electronic microscopy. Three surface ridges and three posteriorly pointing prongs are shown. Such a structure is common on the body, fins, and tail, although scales on the head have a different shape. Courtesy of George Lauder.

mailed the piece to Cornell University, which had a state-of-the-art micro-CT scanner that could scan with two-micron resolution, a fiftieth the width of a human hair. The size of a single shark scale is about 100 times larger than this minimum resolution, suggesting that they would obtain enough data that they could interpolate points in between if necessary. The shark scale was L-shaped and had ridges on its top (Fig. 4.3). It had a long neck with an enlarged base, allowing it to anchor into the soft skin without the use of glue. They used software to create a mesh of the shark scale, a series of points in space that made up an outline of the scale. The shark scale was then digitized into a version that could be 3D–printed. Using software, George and Li placed the scales in a linear array on a flat sheet, as if planting trees in an orchard.

Building a complex shape like a shark scale showcases the contribution that 3D printing can make to biology. The 3D printer created a shark scale layer by layer using a technique called additive manufacturing. The scales were printed on a 3D printer with two inkjet heads that could print both stiff plastic for the scales and a soft rubberlike material for the sharkskin. First the skin was printed, then the base of the shark scale appeared as an island of material within a sea of the printed skin. The printer then injected material for the neck of the shark scale. When it came to the overhang of the L-shaped scale, George and Li were encountering one of the fundamental challenges of 3D printing.

In nature, the shark scale grows like a human tooth, growing inside the body and then erupting through the skin, fully formed. Old worn scales are shed to make room. The new scales develop inside the shark's body, bathed in a sea of chemicals that allows them to grow properly in all directions. In comparison, printing is accomplished layer by layer, and each layer must have its weight supported by something below it. A 3D printer thus has difficulty printing arches and other overhanging structures. Below the overhang in each shark scale, George and Li printed small vertical support columns. When the skin was complete, with its several hundred scales, they used a powerful water jet to break the columns away. Permitting the columns to be washed away is an important part of the design of the 3D–printed part. The columns are generally time-consuming to remove by hand, and since George and Li had hundreds of scales, they had to design the supports to be as small as possible while still supporting the overhang.

They printed a series of prototypes, but in the one that had the acceptable resolution, the artificial shark scales (Fig. 4.4) were 1.5 mm in length, nearly 100 times larger than the real scale. They glued sheets of shark scales to plastic shim stock, sheets printed at set thicknesses used to fill in small gaps during carpentry or building construction.

George put these small shark scale flags in the water tunnel, where they were undulated back and forth by the robotic flapper. He found that the flags increased the speed of the flapper by 6 percent over a

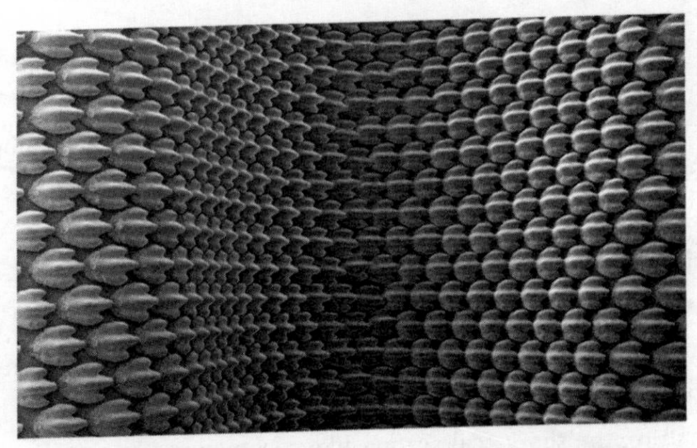

FIGURE 4.4. SEM images of the fabricated synthetic sharkskin membranes. Rigid denticles were fabricated on a flexible substrate membrane using 3D printing technology. Note the changes in spacing among the denticles in the convex and concave portions of the curved membrane, and how denticles overlap each other in the concave region. The scales are 1.5 mm in length. Courtesy of George Lauder.

flapping foil without scales. The leading-edge vortex that he observed with the biological shark scales was also present. The presence of the vortex for the artificial shark scale flapper was reassuring, and suggested that he had replicated the sharkskin accurately.

George and Li's work is only the first step in understanding how shark scales can improve swimming speed and reduce energy use. Just as there are a variety of different weave patterns in fabrics, there are a variety of patterns of scales on sharks. George and Li only tested one scale size and arrangement. In reality, shark scales vary in size from 0.2 mm to 1 mm in size, across both species and position on the shark. The arrangement can be in diamond, honeycomb, and other patterns. As 3D printing decreases in cost further, scientists will be able to answer questions on the function of different sizes, shapes, and arrangements.

While George's work was focused on the streamlining ability of shark scales, it may have other functions. One of them is preventing biofouling, which, as we saw earlier, is the accumulation of algae, barnacles, and other living organisms that eventually coat the undersides of ships and other underwater equipment. A University of Florida chemist Anthony Brennan noticed that algae and barnacles do not grow on sharks and surmised that it was the rough surface of the shark that was responsible. Many anti-fouling surfaces are based on similar principles. For a cell to attach properly to a surface, it must make firm contact, just as we press a piece of adhesive tape to make it stick. Anthony created small bumps in a silicon mold that he used to cast textured surfaces in silicone. Exposure to green algae showed 53 percent less fouling. Synthetic surfaces like Anthony's might be used not just on ships, but potentially also in hospitals, where transfer of bacteria on door handles and other commonly used equipment is one cause of hospital-borne illness.

We have seen that animals can experience benefits by displaying small structures on their surface. Much of the way they do this is by changing the flow of fluids around them. For the shark, we saw that it is the coupling between the body movement and the surface material that reduces drag. In our next chapter, we will learn how animals can take advantage of such movements in ways that permit nearly perpetual motion, that is, motion without the expenditure of energy. We will learn how humans can walk a kilometer on just two teaspoons of sugar, and how dead fish can hold their position in flowing streams. These feats are accomplished by energy capture and storage, concepts that have are just beginning to be implemented into the design of exoskeletons.

CHAPTER 5

Dead Fish Swimming

After seven minutes of walking on the treadmill, North Carolina State University mechanical engineer Greg Sawicki took the lightweight exoskeleton off his ankles. Greg and his student looked at Greg's oxygen consumption over his walk with the brace, searching for a sign that their device was working. The exoskeleton was composed of a carbon-fiber ankle brace that was calf-high like a boot. A wire and a spring ran up the back of his leg like a seam. Flexing the ankle caused the spring to stretch rather than the Achilles tendon. The whole device looked like a cross between an ankle orthopedic brace and a car suspension system. Greg had built the exoskeleton to solve an age-old problem, one that many thought was impossible: to improve the efficiency of human walking.

Twelve years ago, Greg had conceived of the idea of building an energy-saving exoskeleton. Since then, he had worked on and off on the project, as he worked through his PhD and postdoctoral positions. Now, after multiple iterations, he had reduced the weight of the brace to half a kilogram, the weight of a loaf of bread. That also mean that the brace was very fragile, and on the edge of failure each time Greg used it. He lightly placed the brace on a table. Taking off an exoskeleton always feels strange. It's like getting out of a swimming pool and feeling unnaturally heavy in just that one body part. One has to limp for a few steps before remembering how to properly support one's weight again. It's a testament to how quickly our bodies grow accustomed to new environments, without our even noticing.

Greg grew up on Long Island. His parents were nurses whose days consisted of twelve-hour shifts, completely on their feet, so when they came home they were dead tired, and soaked their feet in warm salt water. An average American with an office job can walk over two miles per day. In that labyrinthine hospital, his parents easily exceeded four miles per day. They weren't alone. In fact, many jobs necessitate staying on one's feet. Employees at hospitals, airports, and factories all have fast-paced occupations and require the mobility given by being on one's feet. That's just the world indoors. In the outdoors, legs are needed even more. A standard marine can travel four miles an hour for eight hours a day, walking a total of 32 miles. Despite the development of tanks and Humvees, there is still no replacement for walking. A steep, rocky mountain or a dense forest is no place for most vehicles. In these terrains, the leg trumps the wheel. The average person uses two to three watts, the power of a cellular phone, per kilogram of body mass to walk. If Greg could reduce this amount, he could make a difference for soldiers, and many other people who walk every day. Those disabled by disease or recovering from injury would walk farther before tiring. Although such technology would likely be banned in professional competition, amateur athletes could use it to improve their performance and beat their own fitness goals. Because of the ubiquity of walking in everyday life, a device producing this effect could make a difference for many people. Besides, Greg had always loved exoskeletons in science fiction, ones that could be worn and would amplify a user's strength or reduce their energy use. His dream came to life when, years ago as a University of Michigan graduate student, he began studying rehabilitation robotics, a field that combines electronics, machines, and wearable devices. His graduate thesis was on a pneumatic exoskeleton that could improve human locomotion, specifically the use of energy during walking. It was an ambitious thesis.

The main problem with trying to improve human walking is that walking uses so little energy already. Walking emulates the pendulum, one of the first simple machines ever discovered, and still known for its low energy budget. A typical pendulum has its *fulcrum*, or point of pivot, on top, as in the pendulum of a grandfather clock. An inverted pendulum (Fig. 5.1) has the fulcrum below, like an old-fashioned windup metronome. Thus, walking is also called the inverted pendulum gait. If a pendulum has a well-greased hinge at the fulcrum, it can swing quite a few times before losing all its energy. These swings are made possible by the transfer between two types of energy, kinetic and gravitational energy. A swinging leg does the same thing.

As I walk forward, my body stores kinetic energy associated with its being in motion. As I place body weight on my leading foot, making it my new supporting leg, my body momentarily slows down. Kinetic energy goes into raising my *center of mass* by a distance of 1.5 inches. My body's center of mass is a point designated by the "crash dummy symbol" in Fig. 5.1. The change in position of my center of mass is so large because it greatly affected by the position of my heavy legs. This rising in center of mass is associated with storage of *gravitational potential energy*, which, like a battery, can store energy for brief periods of time. The amount of energy stored is the product of my body weight and the height increase. As my supporting leg pushes off, my center of mass falls downward again a distance of 1.5 inches. Gravitational energy is transferred smoothly into kinetic energy as my body picks up speed again. If I were to track my center of mass as I walk, it would be traveling up and down like a roller coaster. The amplitude of the peaks would be three inches and the wavelength equal to my stride length. Just like a roller coaster trackway, energy is stored at the top of

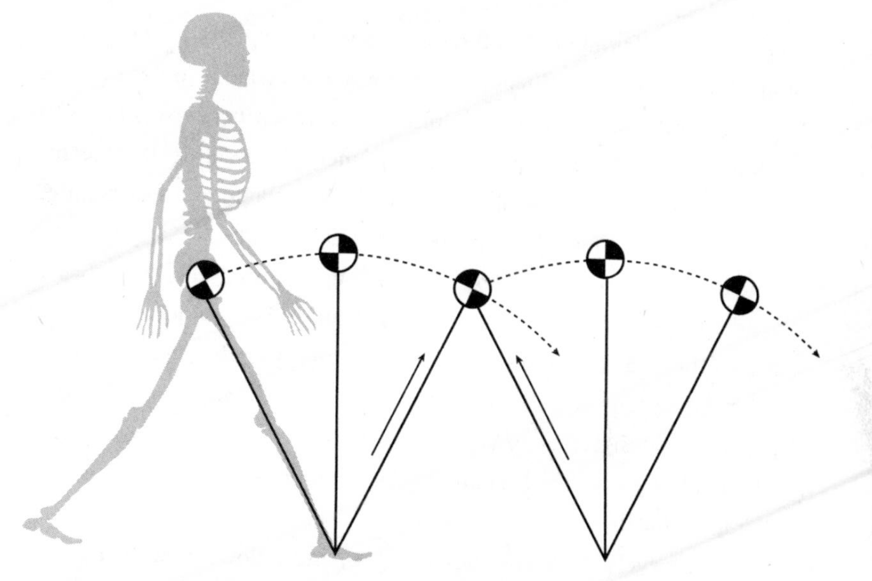

FIGURE 5.1. Human walking in an inverted pendulum gait. As the person's weight is transferred over the supporting foot, the leg acts like a pendulum. The center of mass naturally rises as the leg becomes perpendicular to the ground. Redrawn from Matthis and Fajen 2013.

each peak, and then expended as the "cars"—representing my center of mass—accelerate downhill. This continuous exchange of energy leads to walking's low energy expenditure over multiple strides.

Although the body is very good at cycling energy, it is not perfect. Collisions with the ground cause energy to be lost, and in response, muscles have to inject energy at certain key moments. Greg's mission was to pinpoint the specific locations in the leg where energy was being injected by the muscles. Once he found the timing and location of the energy expenditure, he could replace those motions with an exoskeleton to reduce energy consumption in human walking. When Greg and other locomotion scientists measure energy, they use units called *joules*. In everyday terms, a joule is the amount of energy needed to pick an apple off the floor and raise it to waist height. For every joule of mechanical work produced, human muscle needs four joules of food. In other words, human muscles, for all their versatility, are only 18 to 26 percent efficient. For the sake of argument, let's call it 25 percent. Because of our muscles' inefficiency, when we expend energy walking, we must eat four times the expected amount of food.

The low efficiency of the body converting food to usable work has a flip side: any device that saves the body from doing one joule of work will save the body four joules of food. This energy savings is strong motivation for designing an exoskeleton to reduce the work done by the body. How to design this exoskeleton was the subject of Greg's doctoral thesis. Could he save the human body energy by replacing leg muscles with a machine?

A walking cycle starts with a heel strike, and follows with the rest of the foot. As soon as the foot is flat on the ground, the ankle flexes, with the shin going toward the toes as the leg starts leaning forward. A push-off occurs, when the ankle extends, and the muscles are contracted to push off the ground. The foot goes airborne and the leg swings through the air. During this process, the calf muscles, some of the largest muscles involved, are contracted. These muscles are not necessarily moving very much, but they generate large forces, and, to do so, expend energy and heat up. The bigger the muscle, the more force and heat it generates, and the more energy it would use during a walking cycle if the muscle were activated. Greg suspected the calf muscles were the primary source of energy use in walking and designed his exoskeleton to replace the activity of the calf muscle.

Greg tethered a pneumatically powered balloon just behind the calf muscles to substitute for the muscle contraction. When balloons inflate, they produce force, just like a muscle bulges when it contracts. When the balloon inflated, the ankle joint flexed open, and push-off occurred. Unfortunately, the calf muscles are quite strong. For a balloon to generate the same strength, it had to be tethered to a large scuba tank, controllable valves, and a lot of electronics, together weighing even more than a person. Clearly, the apparatus was far too heavy to become an autonomous exoskeleton. Greg punted, deciding to use the device as a tool to study human locomotion. He hoped that when the device was strapped on the leg, the pneumatics would replace the work of the calf muscles. Experimental subjects strapped on the pneumatic and walked on treadmills for minutes while masks measured their oxygen consumption. This measurement of oxygen could be used to calculate the average energy expenditure over several cycles. For every joule of mechanical energy delivered during the walking cycle, the apparatus saved the user about 1.6 joules of food energy input.

The next step came through the use of technology. In Japan in the early 2000s, scientists had started using ultrasound to look at how muscles move beneath the skin. Ultrasound uses low-energy waves to see through muscle tissue—for example to look at babies in the womb. The computers that controlled the work with ultrasound were improving, and the devices were getting smaller and smaller. The latest ultrasonic probe was the size of a pack of gum. Greg used athletic bandages to strap the probe to his own calf. When he walked, the probe measured how much his calf muscles were stretching. In addition, his oxygen measurements measured how much average energy was used over several cycles. With this instrument, Greg could better pinpoint when and where in the body mechanical energy was being added and subtracted within each walking cycle.

Greg was curious to link the calf muscles' movements with his own oxygen consumption measurements during the walking cycle. He noticed that during much of the walking cycle, the calf muscles appeared motionless. The only phenomenon that was visible was the slow stretch of the Achilles tendon followed by a rapid recoil when the foot pushed off to propel the body forward. The Achilles tendon gives us the spring in our step. It is a biological spring that get stretched out during the rocker motion of the leg as it rotates over the foot's solid base of support. The Achilles stretch is stored up as elastic energy and then returned in a powerful burst, like a catapult, as we unload our leg to the next step. Greg wondered if the part of the push-off that came from the recoil of the Achilles explained why the exoskeleton with a pneumatic muscle wasn't as good as expected. Tendons don't use food energy, only muscles do. Nevertheless, muscles must be using some fuel to hold onto the tendons as they stretch, he thought. Greg dug into the literature, looking into the last decade of studies published about the lesser known functions of muscle.

Muscles have an enormous number of tasks to do in locomotion. Most obviously, they must contract and shorten to do work. When we pick up a bag of groceries, our buttock and thigh muscles contract to allow us to stand up. In this case, the muscle is clearly shortening and doing work by lifting our bodies upward. But even when there is little obvious motion, our muscles can be active. For instance, when we land from a jump, our muscles act like brakes to absorb energy. Some of this energy is dissipated in our soft organs and fat, but much of it is dissipated by muscles contracting and so generating heat. All muscle contractions generate heat, some more than others. A muscle shortening generates the most heat. Second is isometric contraction, where a muscle maintains a constant length.

Muscle can also change function depending on the circumstances. Brown University biologist Tom Roberts in 1997 implanted *strain gauges* into turkey leg muscle to measure the applied forces when turkeys were running. He found that the muscles contracted as expected to give the turkeys an extra boost when running uphill. However, when running on level ground, the muscles took on a different role, acting like a *clutch*, resisting stretch and instead passing it to other structures in line. Resistance of stretch is necessary because the leg muscles are attached to tendons. For the tendon to store and recover energy effectively, it must be attached to a clutch that can react accordingly, locking out while the tendon stretches and recoils.

A similar principle is used in cars: instead of tendons and muscles, a mechanical gear engages or disengages the driving shaft from the driven shaft.

Greg realized that the human calf muscles were acting just like the leg muscles in Tom Roberts' turkeys. In the human leg, the Achilles tendon is attached to the calf muscles. As the Achilles tendon stretches and recoils, the calf muscles try to behave like a clutch, but they have to expend energy to do so. For example, if I hold my arm out, I must continually feed energy into my arm to hold this positon. Small muscle fibers within my arm are quickly tensing and releasing, and losing energy as they do so, in order to hold this position. If I hold it long enough, my arm starts shaking from the effort. A mechanical system such as a table does not expend energy to hold itself in a fixed posture. Instead, its chemical bonds prevent it from moving. The chemical bonds get stretched slightly, and then hold station in their stretched posture. The table does not heat up from the strain, while our muscles do.

Near the end of his time in graduate school, Greg began focusing his efforts on a new exoskeleton that would replace the Achilles tendon with a spring. He and another graduate student at Michigan, Steve Collins, began fiddling with aircraft aluminum cable and a porch door spring to make an artificial tendon out of a cable-clutch. The rest of the device was metal scrap parts, making it too large to walk with, but realistic enough to teach them about how to make it work.

After Greg graduated, he and Steve continued to communicate about designs to improve the exoskeleton. When Greg started his faculty position at North Carolina State University in 2009, his graduate student Bruce Wiggin began working on the project again, based on new concept drawings sent over by Steve. They used springs of varying stiffness and allowed them to be loose during the swing phase, and then taut when the foot struck the ground. They found that if the spring was too stiff, it would lock up the person's joint; too soft, and it did not store energy. In between there was a sweet spot.

Now that they had the spring right, it was time to see how much energy they could save. They needed one more addition to the device so that it could be worn on a human leg. It had to be activated when the Achilles tendon was stretched, but otherwise deactivated to facilitate the swing of the leg. They built a clutch, similar to the one used in cars to switch gears. The clutch was housed in a rectangular box the size of your hand, and was far too heavy to be worn on the leg. Steve had a good eye for shrinking mechanical devices, and he helped turn the clutch into a six-gram device, lighter than a pocketknife, as shown in Figure 5.2B. Greg and his student built a pair of clutches and recruited volunteers to wear them. After about seven minutes of walking with the device (Fig. 5.2 A–B), the testers had consumed 7 percent less oxygen, and so 7 percent less energy. This energy savings was significant. Moreover, unlike Greg's previous pneumatic device, the device was lightweight, wearable, and required no batteries or wall outlets.

Wearable devices in recent years had failed because any energy savings for the wearer would be overpowered by the energy costs of carrying the device. A net savings of 7 percent meant a marine wearing the exoskeleton could increase 32 miles of walking per day to nearly 35 miles per day, for free. People rehabilitating from injuries could walk a little bit longer each day, helping them progress faster toward their recovery goals. Greg believes that the device, which is custom-built to fit a person's ankle, can be on the market within a decade.

Improving human locomotion could also have negative consequences. Wearing the exoskeleton long-term might cause calf muscles to atrophy due to decreased use. As a result, Greg recommends that the device only be worn occasionally, and certainly not continuously. Further, the reduced activity of the calf muscles with the exoskeleton means that the muscles are relaxed throughout the walking cycle. This state causes them to be stretched to a greater degree than they would be when normally activated and used as a clutch. In the long run, this additional stretching could cause micro damage to the muscle fibers.

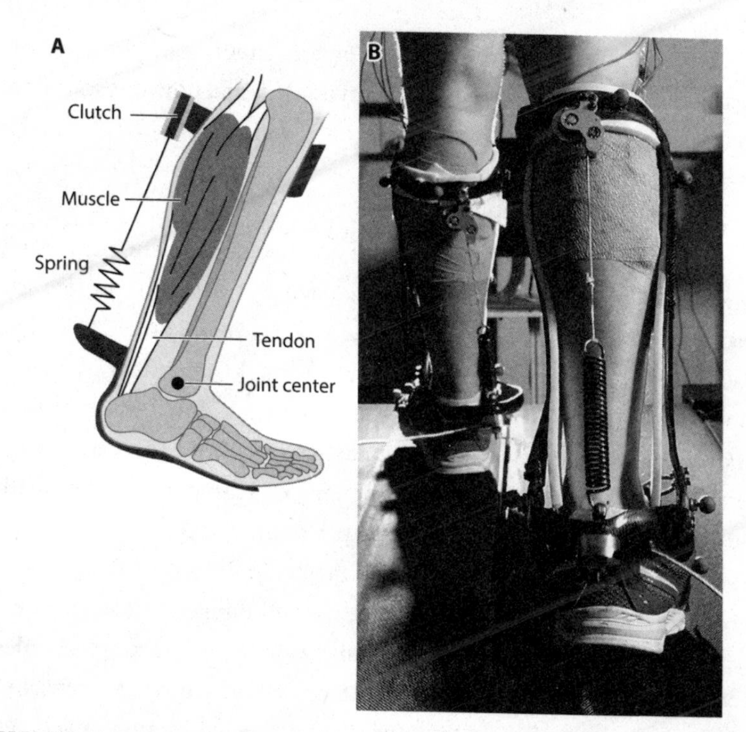

FIGURE 5.2. A wearable exoskeleton that lowers the energy cost of human walking. A. Schematic of exoskeleton relative to calf muscle and Achilles tendon. B. Participant walking with the exoskeleton. It has no electronics, but instead uses a ratchet and pawl that engage the spring when the foot is on the ground and disengage it when the foot is in the air. Courtesy of Steve Collins.

If these challenges are worked out, Greg's approach could be used to replace or augment other parts of the body. His dream would be to design an exoskeleton that would encase the entire leg rather than just the ankle. The exoskeleton would consist of a series of clutches that would lock up one after another during locomotion, saving energy in all parts of the leg. Theoretically, there might one day be a device that would allow one to walk without the use of any food energy at all, except to get going.

The idea of walking without any energy input is not new. In our next story, we will see how this idea originated, and how it inspired the construction of robots that can perpetually walk downhill.

* * *

A modest-looking wooden ramp sat in mechanical engineer Andy Ruina's lab at Cornell University. It looked like a seesaw in low tilt, with a length of 16 feet, one end just a foot above the other. The ramp was actually a power source, where just a small component of gravity could drive a toddler-sized legged robot to walk down the ramp unaided. Like the balance beam in gymnastics, it looked easy, yet most robots failed, falling off the side of the ramp, or crumpling in place with the sound of clanking metal. Next to the ramp was a bin of discarded robot parts where failed designs became scrap. The next ten years of testing on this ramp would lead to the Cornell Ranger, one of the most energy-saving walking robots ever built, capable of walking 40 miles around a running track, fueled by just five cents' worth of electricity. In this chapter, we discuss the pursuit of perpetual motion: machines and animals capable of moving on land and underwater with as little energy as possible.

In the summer of 2001, it was hot in Ithaca. Mechanical engineering undergraduate Steve Collins was just returning from academic leave, where he was deep-frying fish as a short-order cook. Throughout the summer, he walked up and down the ramp in Andy's lab hundreds of times. Plodding down the ramp next to him was the lab's latest walking robot, a headless and torso-less pair of legs and two swinging arms, yet with a surprisingly human gait. On a good day, a single push by Steve sent the robot walking the 20 steps down the ramp on its own. It made loud clanking sounds when its knees swung shut, and thumping sounds when its wooden feet hit the ground. The robot's interactions with the ground were far too complicated to predict using a computer. Therefore, Steve spent the entire summer tinkering, making small adjustments to get the robot to walk straight. This particular robot was called crude by some—it was a prime example of an *under-actuated system*, designed with a minimum of actuators or motors. Steve and Andy were driven by a hunch, a gut feeling that much of the field of robotics was on the wrong track toward building energy-saving walking robots.

Just the year before, following the culmination of 15 years of research, Honda released Asimo, the most sophisticated walking robot in the world. Now, over 40 copies of Asimo are all over the world, shaking hands with world leaders, dancing on stage, and playing soccer. At over one million dollars each, they are excellent robotic ambassadors, robot celebrities even. But at the time of Steve's research, Asimo had a problem: its battery life was a maximum of 40 minutes. This was because Asimo used two kilowatts of power to walk, more than 20 times as much as a human. How could Asimo be so sophisticated, yet so much less economical than a human?

Asimo and most other walking robots have one thing in common: they are designed based on a "kinematic obsession" with walking. A walker's *kinematics* consist of the time course of its joint angles: the progression of how its ankle, knee, and hip joints change with time. People believe that a bipedal robot only walks correctly if its kinematics match that of a human. To enforce correct motion of each of the joint angles, a motor is commonly placed at each of the robot's joints. By timing the motion of each of these motors, the ankles, knees, and hips can be flexed in the appropriate sequence. Only then is walking replicated. That is the idea, at least, yet the dream has yet to be achieved. At the time of Steve's research, Asimo had 57 individual motors, and a sophisticated computer system to control them. But when these motors were timed to rotate, Asimo developed an unnatural gait, one that is very unlike a human's. It walked on hunched knees and looked like a thief that is trying to avoid being caught.

Andy, who was Steve's advisor, strongly opposed the kinematic obsession of walking. Instead, he thought robots should attempt to perform *passive dynamic walking*, whose goal is to take advantage of the human body's natural ability to walk. The key idea is that when we walk, most of our muscles are actually turned off. Andy took this idea to its very limits by designing robotic puppets, those with all the joints and segments of the leg, but not a single motor. It was a minimal approach, and many did not believe it would work. But it did take advantage of a single important rule for walking that Asimo did not. This rule was discovered by Newton two hundred years ago, and then written into his *laws of motion*.

Newton stated that a body in motion tends to stay in motion. In particular, horizontal motion can last forever without the input of energy. Consider a wheel that is sent rolling down a road. If the road is perfectly flat and straight, and the wheel perfectly rigid, it simply rolls forever. This is exactly the principle that bowling alleys use to get the balls rolling the 50 feet down the alley. Both the hardwood floor and the bowling ball are quite rigid objects, reducing energy losses manifested as losses in ball speed. Indeed, a wheel is the perfect machine for moving between two points on a horizontal surface. Riding a bicycle is four times more energy-saving than walking. Still, the leg has several properties that make it as wheel-like as possible.

At first glance, a wheel and a human leg do not look very much alike. But imagine a wagon wheel with many spokes, and the human leg as simply one of the spokes. When we put our leg down, we bear our weight on the ground, just as a wagon spoke does. As we take a step, our foot rolls upon the ground like a wheel, as weight is shifted from heel to toe. To save energy, our center of mass should translate horizontally without bobbing up and down. If it can do so, it will be as energy-saving as possible, achieving a total absence of energy loss. This is how wheels can roll for so long on flat, rigid surfaces. Their center of mass, or the center of the wheel, simply translates forward.

Newton showed us that zero energy use is possible for a wheel rolling across a flat surface. But moving on legs is another story because losses are accrued in a number of instances during the stride. When we transfer our weight from one foot to the other, our center of mass falls slightly, about an inch or two. When the center of mass falls, the gravitational energy is recaptured as much as possible. For humans, energy is stored in elastic structures such as the Achilles tendon and in the soft elastic structures such as the fatty pads on the bottoms of our feet. These structures act like coiled springs that capture energy and return them to us the next stride. They give us the "spring in our step."

How close are we to the perfect locomotion of a wheel? Just consider the facts. We take about 7500 steps per day. In our lifetime, the average human will have walked over 36,000 miles, about 1.5 times around the earth's equator. Each kilometer involves over 3280 steps, yet is fueled by the energy of just two teaspoons of sugar. Indeed, human walking is quite economical. But it is not because a human is doing it. In fact, even simple wooden walking toys can walk effortlessly downhill.

* * *

If passive dynamic walking had a mascot, it would be a thumb-sized wooden penguin called the Wilson Walkie. It was invented by John Wilson in 1938, during the Great Depression, in a small town called Watsontown, Pennsylvania. At the time, plastic and battery-operated toys had overtaken the toy market. Wooden toys were on their way out. But the Wilson Walkie was fascinating because it was deceptively simple. A cone-shaped body sat atop two jointed wooden legs, slack like Pinocchio's. To play with it, one placed the penguin at the top of a shallow ramp similar to the one in Andy's lab, and then lightly tapped it on its back with a finger. If pushed just right, the toy would then walk down the ramp, just like a real person. The toy required careful design to reduce friction and enable the transfer of energy from step to step. The Walkie fad lasted a decade before it was overtaken by other toys.

Nearly forty years later, in the 1980s, the toy began to inspire mechanical engineers, including Simon Fraser University aeronautical engineer Tad McGeer. McGeer was inspired to copy the toy design and expand it into a larger metal version. The larger Walkie had four legs in a row that moved in pairs, first the inner pair, then the outer pair. By tinkering, he realized his larger Walkie could also successfully walk without much intervention. One of the big contributions McGeer made was adjusting the knee so that when the robot extended the knee, it would not get locked in place. However, the robot still did not resemble a human walking because it relied on four legs. In Andy's lab, it was up to Scott Collins and Martijn Wisse, a visiting student from Netherlands, to reduce the number of legs to two.

Reducing the number from four to two is difficult because of the reduction in the robot's stability. A four-legged robot has a pair of legs on the ground at all times, which means that it can only fall backward or forward. Walking thus reduces to a two-dimensional problem. A bipedal robot has only one foot on the ground at a time, which gives it the problem of spinning on one foot like an ice skater when it steps. Any spin causes the robot to veer away from a straight path, or worse, topple over entirely.

To overcome the instability of standing on one foot, Steve worked with a robot that Martijn had constructed the previous summer. The most difficult parts were the robot's feet. When a foot hits the ground, its material properties and shape dictate where the robot travels next. The worst design for a foot would be to make it hard and angular, like the edge of dice. When we roll dice on a table, each die usually bounces off one of its edges, making it difficult to predict which side it will land on. Steve used springs to implement a compliant heel for the robot so that when the heel hit the ground, the contact time would be extended, enabling the robot to remain in control of its motion.

Next was the shape of the foot itself. Steve wanted to emulate a wheel as much as possible, to ease the weight transfer between the heel and toe. After a few designs, he settled upon a foot composed of two plywood rails, each shaped like biscotti. The rails themselves were coated with rubber to increase friction and soften impact with the ground. If the inner rail were made taller than the outer one, the robot would naturally rock side to side as it walked, like a penguin. This rocking gave room for the other foot to swing into place, since both legs were of the same length. It also helped to maintain stability. With only one foot on the ground, the center of mass had to fall within the foot's outline, lest the robot topple over.

Now that he was finished with the feet, Steve continued with the rest of the robot. He used a single string to hang the robot from the ceiling like a puppet. As he pulled back a leg and then released it, the robot spun around its other foot like a top. This spinning was a manifestation of *conservation of angular momentum*, and it occurs not just for robots but for humans or any bipedal animals. The leg is so heavy that when it swings, the body naturally wants to spin in the opposite direction to conserve angular momentum. Similarly, if you extend your arms while spinning on a wheeled chair, your rate of spin will decrease due to conservation of angular momentum.

When we walk, we usually counter the leg swing by swinging the arm on the opposite side. It's possible to walk with your arms strapped to your sides, but it requires muscular effort to resist the rotation of the pelvis and shoulder. As a result, energy use is 3 percent less if arms are swung normally than if they are held behind the back, and 9 percent less than if held in front of the body. So, in a sense, we walk with both our legs and our arms. Steve built a pulley system that pulled on a counter-swinging metal arm, which rotated forward and outward as the opposing leg was swung forward. These arms helped reduce rotation of the body and also helped stabilize the side-to-side lean. Steve's robot with its counter-swinging arms is shown in Figure 5.3.

For much of the summer, Steve watched the robot closely to see how it failed as it walked down the ramp. When it failed he gave the robot a corresponding tweak. The next time down the ramp, it would fall in a different way, so that would again require yet another tweak.

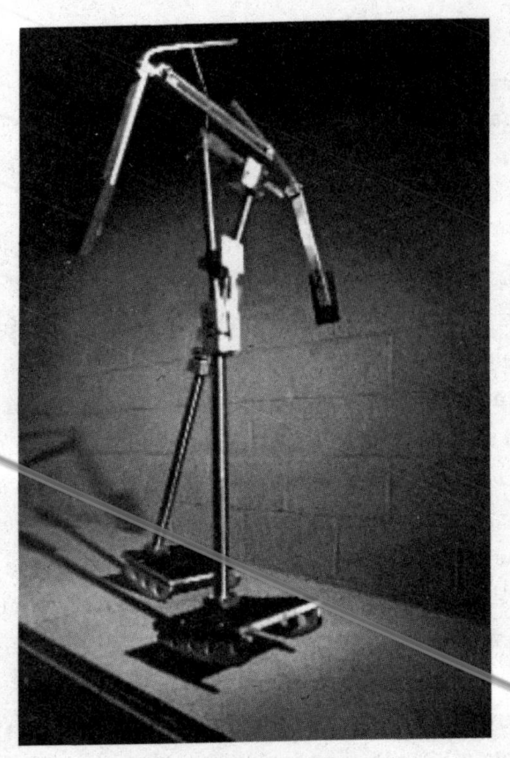

FIGURE 5.3. A two-legged passive-dynamic walking robot. The counter-swinging arms are attached rigidly to their opposing legs. The right arm swings forward and out as the left leg swings forward. The robot stably walks down inclines using only gravity as a driving force. Courtesy of Steve Collins.

This was hard work, but also addicting because as Steve tweaked, he found that the robot performed better, getting closer to the end of the ramp. This manual tweaking was the fastest path to getting a working robot. Previous scientists who worked with four-legged walkers could perform optimization on a computer, since the robot was walking in two dimensions. However, a three-dimensional robot could fall forward, backward, sideways, or spin, and handling this was beyond what a computer could do at the time. Moreover, while the equations of motion were known, the boundary conditions were not. Solid-solid interactions such as collisions, rolling, and friction are still not well enough understood to program into a computer. Thus, just like fixing a bike, it took old-fashioned tinkering to get this robot to work.

On a good day and with Steve's accumulated experience launching the robot down the ramp, it walked steadily 80 percent of the time. Because the robot had no control systems except for its self-stabilizing passive dynamics, it was quite sensitive to its initial conditions. When successful, it walked down the incline at 0.5 m/s, less than a third the walking speed of a human. However, the robot's energy use was a mere 1.3 watts; in comparison, the average human at rest uses 100 watts of power. A 130-kilogram version of the walker would use 34 watts, still nearly 60 times less than the Honda Asimo. Passive dynamics gave the promise of under-actuated robots, those with few motors, in contrast to Honda Asimo's hundreds of motors.

Now, more than a decade later, those who support passive dynamics are coming to a compromise. The problem with passive dynamic walkers is that they fall too easily. That was why Steve had to spend an entire summer trying to adjust his robot's feet, and even when they were adjusted, had to stand within just a foot of his robot as it was walking. Now, even Steve's advisor Andy agrees that the most important design criterion for walking robots is to avoid falling. That means adding sensors and motors to modulate the gait in real time. Andy believes that robots of the future will emulate the energy savings and elegant walking style of passive dynamic walkers, but also rely on a number of actuators to prevent falls. Such robots will be a compromise between the complexity of Honda's Asimo and the simplicity of the Wilson Walkie.

In our previous two stories, we learned how to make humans and robots walk with greater fuel economy. In Greg's prosthetic, a device took the place of a muscle to reduce the energy consumption of a muscle when the muscle acted like a clutch. In Steve's passive dynamic walker, energy was transferred from potential to kinetic energy just like in the human leg. By transferring energy, animals and machines can increase their fuel economy. In our next story, we will see how fish take advantage of the energy around them to swim efficiently.

* * *

Harvard graduate student Jimmy Liao gently lowered the dead trout into the gushing river simulated by the water tunnel. He shook his head as he did. Common sense told him that this was going to be a fruitless test. But all the scientific evidence from the last two years of his doctoral thesis was nagging at him, and he had finally given in. He simply had to answer the question, can a dead fish swim? Many thought that fish swim in water, but Jimmy would show that in certain circumstances, water can swim a fish.

Jimmy was tan from years of trout fishing, which he'd been doing since he was six. Most of the boys at Jimmy's school played football. Instead, every day after school Jimmy and his friends sat on a riverbank and fished for trout in the Sawmill River, whose lazy currents flowed through Pleasantville, the suburb where Jimmy grew up. Those afternoons, the trout became Jimmy's friends. He learned what lures they like, what weather makes them active, and how to move the line to draw a chase. The most important trick he learned was where trout rest. Trout hid behind the downstream face of a rock to avoid fighting the currents. This is where Jimmy and his friends released their lines and relaxed after a day of school.

When Jimmy began graduate school at George Lauder's lab in Harvard's Museum of Comparative Zoology, the study of fish was undergoing a transformation. The computer age beginning in the 1980s was impacting every field of engineering, especially fluid mechanics. Researchers had previously visualized the motion of the fluid by seeding into the flow small bright particles called *tracers*, named so because they traced out the motion of the fluid, as if drawn with chalk. The use of tracers was critical in the development and tuning of modern airfoils by aerodynamicists such as Ludwig von Prandtl, the father of modern aerodynamics. In the 1970s, the invention of laser light enabled only a thin plane of the fluid to be visualized. This technique provided nice pictures, but could not yet measure flow speeds. The computer revolution of the 1980s enabled the use of a computer to record the position of the particles and calculate their speeds. This technique, called *particle image velocimetry* (PIV), became the bread-and-butter of most aerospace companies, and in the 1990s Jimmy's doctoral advisor George Lauder was the first to use PIV to study fish.

On the heels of this influential piece of technology, Jimmy Liao arrived at graduate school. For his doctoral thesis, Jimmy proposed to do experiments with trout, a flashback to his childhood fishing days. Specifically, he wanted to find out what benefit trout get from hiding behind obstacles. He hypothesized that trout were *drafting*, a technique bicycle racers use to save energy, in which a rider follows closely behind another rider. In drafting, the distance between the riders is key. Too far, and the aerodynamic shield provided by the leading rider is insufficient. Even farther, and one becomes buffeted by the wake of the leading rider. Jimmy wondered, what distance did trout choose to stay behind a rock?

Jimmy ordered trout from nearby trout-breeding farms. They arrived in the mail and he raised them in large eight-foot-diameter drums with constantly circulating water. On the days that he did experiments, Jimmy netted the fish and brought them over to the flow tunnel, a fish tank with a powerful pump at one end. The pump caused the water to flow down the length of the tank, simulating a fast-moving stream. In front of each pump was an evenly spaced plastic grid used to filter out eddies that would disturb the trout. After water passed through the grid, it was *laminar*, or smoothly flowing.

Jimmy was friends with David Beal, a graduate student in ocean engineering at MIT, just down the street from Harvard. David was in

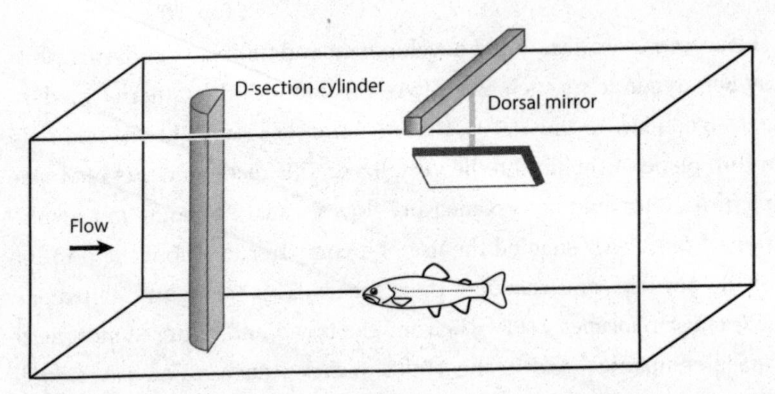

FIGURE 5.4. Trout swimming behind a D-shaped cylinder in a flow tank. The arrow represents the direction of water flow. As fluids flow past the cylinder, it sheds vortices that influence the motion of the fish. Redrawn from a figure by Jimmy Liao.

a lab that was working on Robotuna II, an update of a robotic tuna. Jimmy would meet him after school, just like when Jimmy was back in Pleasantville, but instead of catching trout, this time he was trying to understand how they swim.

Jimmy wanted to understand how fish would react to obstacles placed in the flow tunnel. At first, he tried mimicking nature by placing sticks and rocks in the tunnel. The problem was that there were too many variables to get right in placing these kinds of oddly shaped obstacles. It was hard to do the same experiment every time. His engineering friend David suggested that they use a D-shaped cylinder, a cylinder that has gone through a buzz saw so its cross section looks like the letter "D," as shown in Figure 5.4. The result was a long D-shaped rod the width of your fist, but long enough to extend from the bottom to the top of the flow tank. Of course, trout in the wild do not swim behind such odd shapes. But the D-shape cylinder was best because they could use it to control the wake precisely. The object was long and thin in order to generate a two-dimensional flow, a flow that looked the same to the fish no matter what height the fish chose to be in the tank. This is important because fish do what they want in the tank. The cylindrical part of the shape has a long history in the field of fluid mechanics. Cylinders are chosen because they are easily described using mathematics. A simpler mathematical description makes calculations of the wake behind the cylinder also easier to describe. A cylinder in a flow of sufficiently high speed generates an oscillatory wake called a Kármán vortex street that looks like the swirls on a tie-dye shirt. You would expect the flow past a cylinder to look a lot like the flow in front of the cylinder, simply fluid continuing on its way. But in fact, a cylinder generates a repeating pattern of swirling vortices. This is an example of a fluid instability, where uniform flow is tripped up by an object and creates an oscillating flow. The D-shape makes the experiment more repeatable because vortices are shed off the cylinder at the corners of the D-shape, and so at the same locations in each trial.

The day that David brought over the D-cylinder, Jimmy and he spent the entire day observing a trout in the water tunnel. The lights in the room were off, and the green laser light sheet was on. The only sound was the gurgling of water from the pump driving the fluid in the flow tank. For hours, the trout would just avoid the obstacle, staying in the corners of the tank. It was late afternoon, and Jimmy and David were tiring, about to give up hope of understanding how trout hide behind boulders. Before bringing the fish back to the lab, Jimmy decided to do one last experiment. The pump whirred loudly as he cranked up the flow in the tank as fast as it would go.

Jimmy and David watched as the trout was sucked into the region behind the cylinder as if a vacuum cleaner had been turned on. This was the moment they had been waiting for. They cheered. Then, even stranger, the trout stayed right in that position, its head stationary, fins pressed against the sides of its body, and its tail waving back and forth like a metronome for minutes at a time (Plate 8). It was as if the fish's body was replaced by a flag. Anyone who has ever watched any aquarium knows that this is odd behavior. Fish usually swim with a combination of all their fins, pectoral, tail, and dorsal, and so usually do not hold their position, but are constantly exploring their surroundings, moving quite randomly. Seeing the fish become a stationary, tail-beating object was odd indeed.

The fish's behavior was also completely contrary to the state of understanding of how fish propulsion works. At the time, the view was that fish push fluid backward to move. This principle had been seen time and time again. It was also true for each of the fish's many fins. The motion of each fin sends a small vortex backward, and the fish is pushed in the opposite direction. How a fish would behave when vortices were thrown onto it was a whole other question. At the time, most research efforts were focused toward seeing how fish behave in either stationary flow or flow of constant speed. Few had considered how a fish reacts to vortices. In fact, observing a fish encountering vortices gives a much more accurate view of how nature appears to fish in fast-moving rivers and even in the waves of the deep ocean.

Jimmy thought about the fish's bizarre motion and decided that it was a new gait. A *gait* in animal locomotion is a pattern of body movement that propels the animal forward. Horses have a walk, trot, pace, canter, and gallop. There had been several classifications for fish gaits already, but none of them fit the slow beat of the fish's gait behind the cylinder. Jimmy decided to call it the Karman gait. But how was the fish creating this strange gait? Jimmy decided to use electromyography, EMG, a technique to measure the electrical activity of muscles. It was first invented in 1773 by Francesco Redi, who discovered that the muscles of the electric ray fish could generate electricity. By measuring the activity of muscles in the fish, Jimmy could begin to understand what the fish was doing with its body during the Karman gait.

The EMG technique involves slipping the tips of wires under the skin of the fish. In this way, the fish's muscles were treated like a battery, providing voltage that can be measured. The fish were put to sleep with MS-222, a substance like powdered sugar that dissolves when mixed with aquarium water. In lower concentrations, it sedates or anesthetizes fish, but if the concentration is too high, the fish dies. Once the fish was anesthetized, small wires the size of acupuncture needles were slipped just underneath the skin. These wires were then connected to an amplifier that magnifies the electrical signal measured between the wires. The fish with wires sticking out of it was then replaced into the flow tunnel and slowly revived. The computer connected to the EMG wires gave loud blips signaling each muscular contraction. As the fish began its Karman gait again, the computer made much fainter sounds. Perhaps something was wrong with Jimmy's setup. But if the setup was correct, it meant that the fish was not using its muscles at all. Then, what was causing the fish to move its tail back and forth?

Jimmy had a hypothesis: the fish was relaxing all its muscles and letting the fluid move its body. The living fish had become a passive flag. In other words, it had become an energy-harvesting device. Instead of using its energy to maintain its position behind the cylinder, it simply relaxed its body and used the fluid to move its body into the correct position. He hypothesized that the series of flow patterns generated by the cylinder generated enough hydrodynamic force to hold the fish in position. It was a wild theory at the time. No one had ever heard of a live fish playing dead and extracting benefits from its surroundings. Jimmy decided to put his theory to the ultimate test. He hated killing fish, so he did this test only once.

Jimmy increased the concentration of MS-222 in the tank, killing the fish. He put the dead fish in the flow tunnel. As soon as it was within range of the wake of the cylinder, it was sucked into the wake and also began undulating its tail. It was like the fish had returned to life! In reality, the dead fish was not swimming, but instead it was being swum by the fluid around it, like a puppet. The puppet master was the vortices surrounding the fish, which could shake the fish's flaccid body back and forth. It showed that if a fish relaxed its body sufficiently, it could harvest the energy around it, and could thus move forward relative to the fluid. It is more than just the way a kite is pushed by the wind. It is as if the kite has a fish's body and is swimming upwind. Jimmy's experiment showed that turbulent motion in fluids is not always a drawback. When we sit in an airplane, we dread turbulence. But if it is a special kind of turbulence, fish can actually exploit it, drawing energy from it to save their own energy.

* * *

Each of the stories in this chapter involved the concept of energy. Greg Sawicki built a device that supplanted the contraction of a muscle, saving energy during human walking. Scott Collins built a walking robot that could harvest the energy from gravity to walk steadily downhill. Jimmy Liao showed that dead fish can swim because they harvest the energy from their surroundings. All of these are methods to get from A to B using less energy. Sometimes, saving energy is not the goal animals have in mind. There are times that animals have to get from A to B no matter what the obstacles. To do so, they have to be resilient against impacts, as we will see in our next chapter.

PLATE 1. A Labrador retriever performs the wet-dog shake, removing a pound of water from its fur in a fraction of a second. The Labrador's loose skin whips back and forth as it shakes, amplifying the motion and increasing the number of drops that are released. All furry mammals tune their shaking frequency to their body size to achieve the centripetal forces necessary to shake dry.

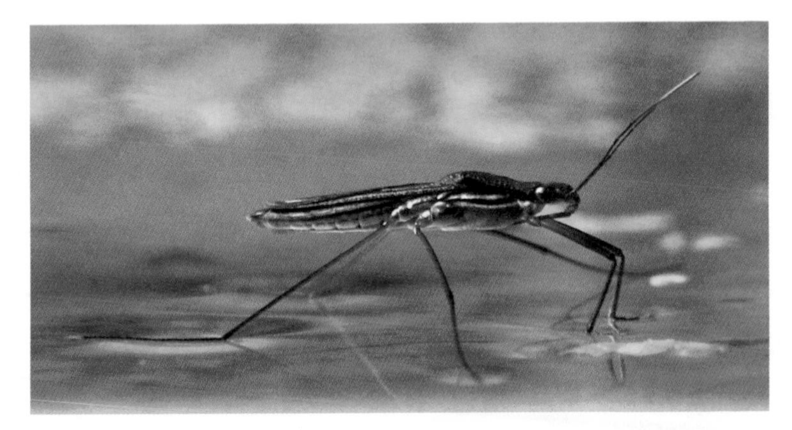

PLATE 2. The water strider *Gerris remigis* is a common insect found on ponds and streams. It propels itself by rowing on the water surface and uses the surface tension of water to support its weight. Its legs are water-repellent, allowing it to skate with low friction across the surface of water.

PLATE 3. Close-up of the legs of a water spider, whose numerous hairs make it energetically costly for the legs to become wet. The silvery sheen indicates that a thin air layer is trapped, an indication of high water-repellency.

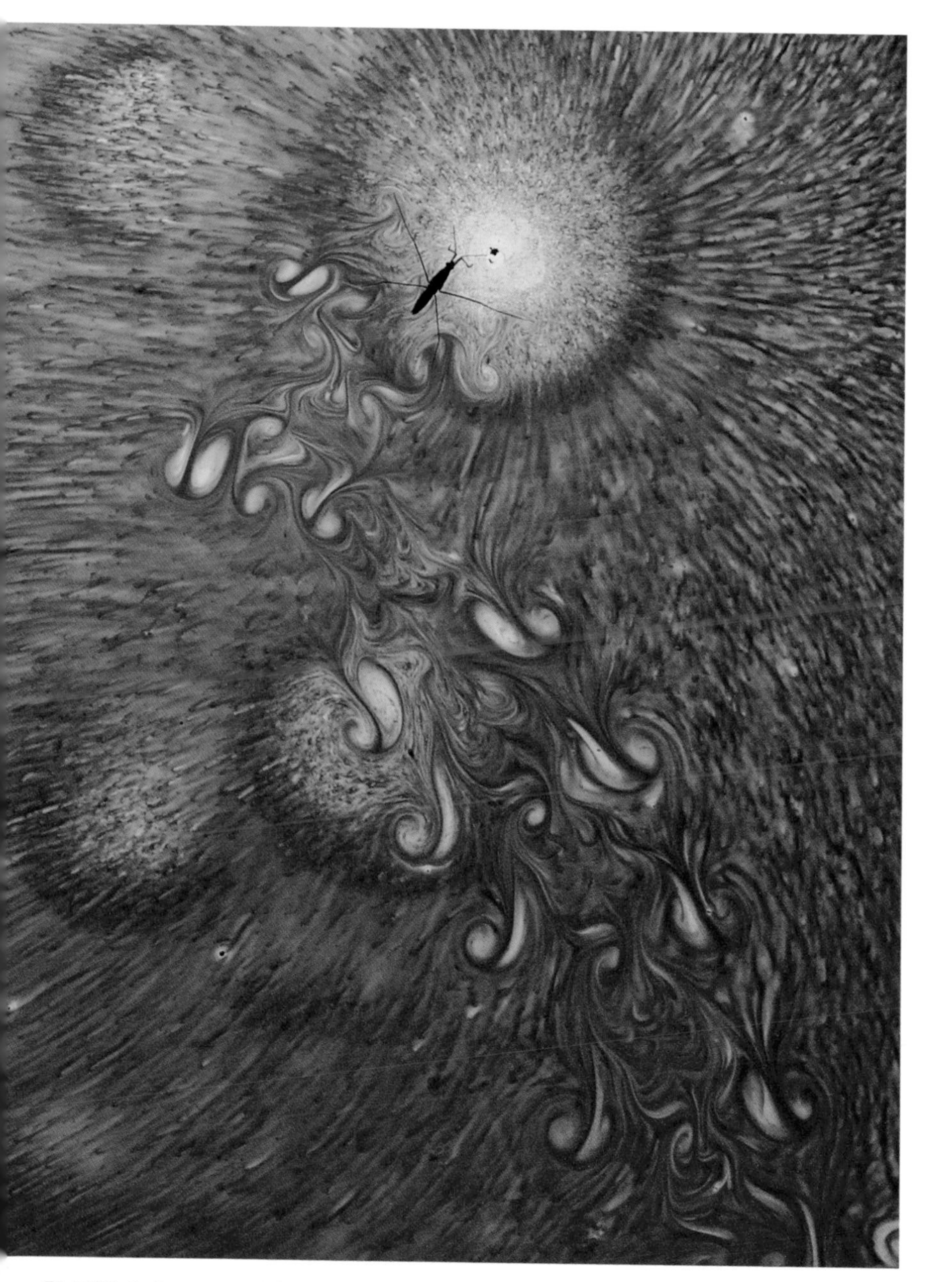

PLATE 4. A water strider rows on the water surface, generating an underwater vortex with each stroke. The vortices are made visible by lighting the water from below and adding the pH-activated dye thymol blue. A particularly large chunk of the dye created the sunburst at the top of the image.

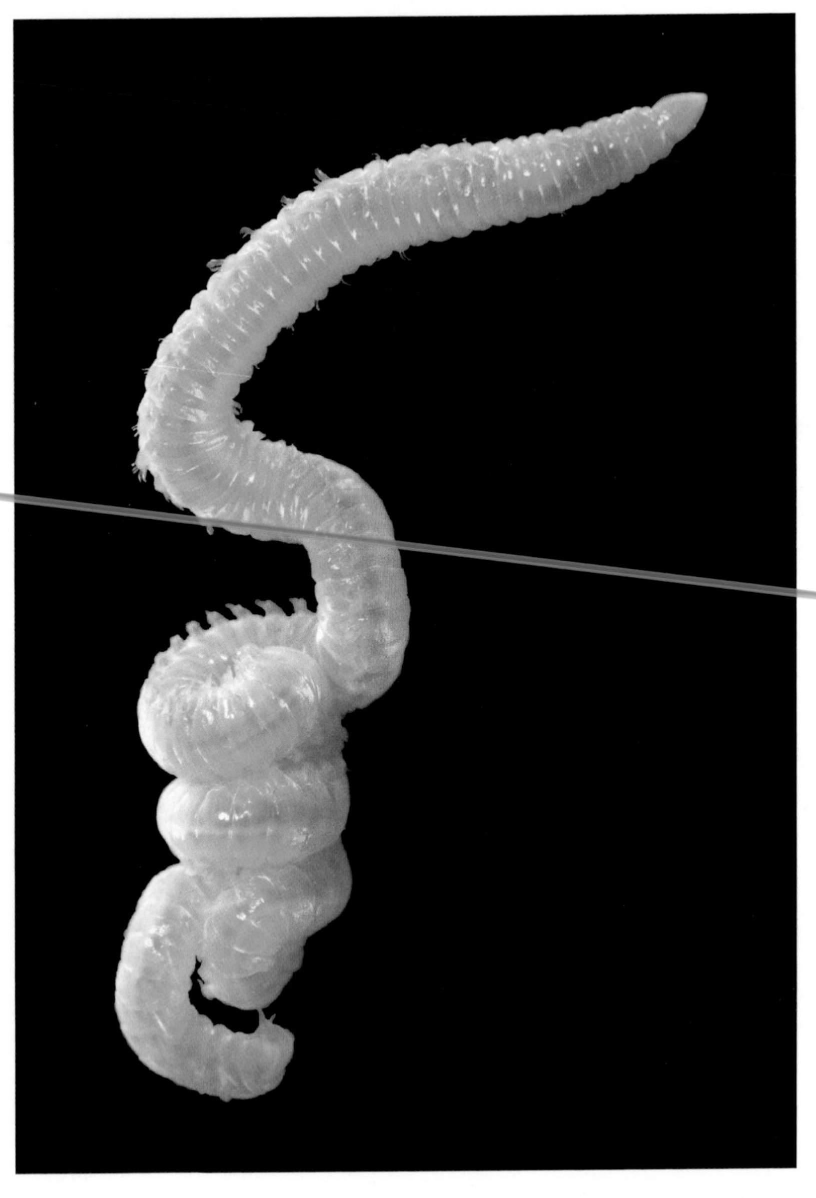

PLATE 5. A marine worm called a polychaete, which translates as "having many bristles." The mud surrounding the worm is quite stiff, but by wiggling its head back and forth, the worm can generate a crack ahead of it, similar to how an axe splits logs. The worm propels itself by generating cracks to weaken the surrounding mud. Courtesy of Kelly Dorgan.

PLATE 6. An elephant has the largest bladder of any land animal. It urinates with a flow rate of five showerheads, releasing 19 liters of urine, about a kitchen garbage can's worth, in an average of 21 seconds. This urination time is conserved among mammals, from a dog to an elephant.

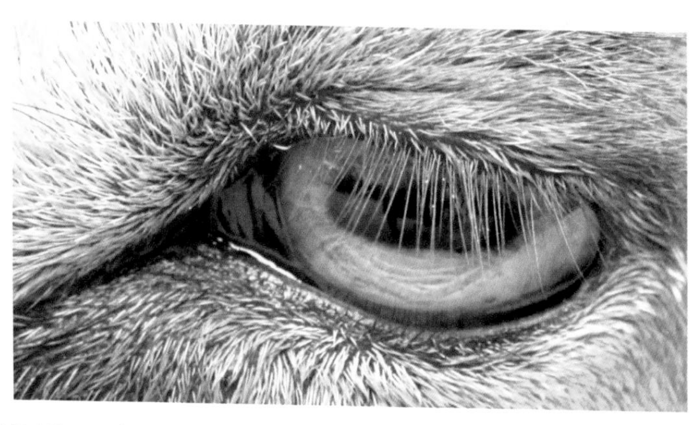

PLATE 7. The eyelashes of a goat. The lashes of a goat and other mammals deflect incoming air as the animal walks forward. With less incoming air, the eye's tear film remains intact longer and intercepts fewer grains of pollen and dust.

PLATE 8. A trout slaloms between vortices, harvesting the energy of the surrounding fluid to reduce its own energy use. This energy transfer is so effective that a dead trout can accomplish the same feat. Courtesy of Jimmy Liao.

PLATE 9. A mosquito flying through a rainstorm recreated in my lab. Although raindrops can weigh fifty times as much as the mosquito, the mosquito escapes unharmed. To survive, the mosquito uses its low mass to its advantage. Because the mosquito is so lightweight, the raindrop does not slow down upon impact, and transmits less force to the mosquito.

PLATE 10. Fire ants, *Solenopsis invicta,* use their legs to link their bodies together, constructing a waterproof raft. The raft can flow like a fluid or bounce like a solid, traits that help it survive impacts with obstacles. The raft floats on water until it reaches shore or emerging vegetation, where the ants will disembark. Courtesy of Anand Varma / National Geographic Creative.

PLATE 11. A cluster of ants slowly pulled apart. The network of ants begins to separate into long chains, similar to hot cheese. The ants' tendency to flow demonstrates their fluid-like properties, which enables them to change the shape of their structures.

PLATE 12. To cross ravines in their path, ants can link their bodies together to build bridges. Weaver ants are shown here, but fire ants and army ants can also perform this feat. Depending on the traffic on the bridge and the need for new foragers, the bridge can expand or contract in width. Courtesy of Chris Reid and Christoph Kurze.

CHAPTER 6

Flying in the Rain

How can mosquitoes fly in the rain? A typical raindrop and a mosquito are about the same size, a couple of millimeters. But a raindrop is spherical, and much denser. Consequently, a falling raindrop weighs about 50 times as much as the mosquito, relatively as much as a pickup truck with respect to a human. Even the smallest drizzle droplet is about the same weight as a mosquito, and travels faster. How can the mosquito survive the impact of one raindrop, let alone the barrage of raindrops during a typical downpour?

The mosquito faces a problem common to many insects. They live in a world where they are completely outnumbered by large, unpredictable obstacles. As a foraging bee flies through a field of flowers, the stalks swing back and forth, blown by the wind in unpredictable ways. When a cockroach runs along the ground in the dark, collisions are unavoidable. This chapter is about the adaptations insects have for dealing with collisions.

To study how mosquitoes survive rainstorms, my graduate student Andrew Dickerson and I took a visit to Atlanta's Centers for Disease Control. When we entered the mosquito breeding chambers, I had the eerie feeling of being watched. A high-pitched humming sound emanated from the cages all around, the combined wingbeats of tens of thousands of mosquitoes waiting to feed. As I leaned closer to the cages, I saw them lunging toward me. I could even feel the slightest breeze created by the flap of their wings, like a million snowflakes falling at once.

In the past, to feed the mosquitoes, the CDC scientists had left rabbits in the mosquito cages. Over the years, a lab technician Paul Howell had come up with an easier method. Paul could often be seen with an arm beneath a black cloth. Upon further inspection, we found him feeding mosquitoes with his own blood. "Mosquitoes prefer to feed in the dark," Paul said. The blood suckers had the ability to bite Paul's arm right through their mesh cages. All he needed to do was place his arm in contact with the side. Paul continued talking to us as his blood was being drained. After a minute he was done, and he removed his hand, covered in thousands of small red points, as if painted by a pen. His body had grown so used to it that he no longer swelled.

Andrew and I returned to Georgia Tech a little disturbed, but happy to have several hundred mosquitoes in a cardboard canister. Such mosquito pickups would become commonplace for us over the next few years. Back in our lab, Andrew had built a rainstorm simulator. Mosquitoes were too small, and raindrops too unpredictable, to ever capture their collision on camera in nature. To increase the odds in our favor, Andrew built an acrylic rain chamber about the height of a shoebox and the width of a fist. A mesh bottom allowed drops to fall through. A hundred mosquitoes buzzed around in the narrow chamber, excited by the smell of Andrew's breath. Suspended above the container was a pump that released drops at a fixed size and speed. With drops coming regularly along the same path, any mosquito that was in the path of the drops was bound to be struck. All we had to do was set up our high-speed camera and wait.

We couldn't see anything with our bare eyes. All we knew was that the onslaught of raindrops did not seem to be affecting the mosquitoes, who continued to buzz in the container. When Andrew captured the

FIGURE 6.1. Two image sequences of a mosquito struck by a raindrop. (Top row) The drop strikes a mosquito on the wings, causing the mosquito to rotate in midair. (Bottom row) A mosquito is pushed downward for a brief instant. Both mosquitoes easily recover and continue flying.

first mosquito getting struck by a raindrop, we had no idea what to expect. He and I both watched eagerly as we slowed down the video (Fig. 6.1, Plate 9). We saw a drop falling frame by frame. When it finally struck the mosquito, the drop compressed slightly, but then it kept going. There was no splash. The mosquito-cum-raindrop fell together for a few centimeters before the mosquito slid off. The drop continued to fall and the mosquito continued its flight, both completely unharmed.

This brief interaction between the drop and the mosquito was the secret to how the mosquitoes survive such an impact. Their interaction is quite different from our experience with rain. If you extend your palm in a rainstorm, drops strike your palm and then shatter into small pieces. In turn, you feel a force on your hand because it is resisting the raindrops' motion. However, the mosquito weighs only 2 percent of a raindrop. It is so lightweight that it does not resist the raindrops' motion. Like a tai chi master, the mosquito simply lets the drop continue unimpeded. Imagine trying to punch a balloon with all your might.

Because the balloon does not resist your fist, you cannot do it damage. In fact, it's impossible to kill a mosquito by swatting at it with one hand. You are simply pushing it along for a ride. The only real way to kill a mosquito is to clap it between two hands or pin it between your hand and the part of you that it is biting.

Rainstorms are not the only source of obstacles. In our next story, we will see how foraging bees deal with crashing into flowers and other plants. Here, the impact is much worse than being struck by a moving raindrop. As someone who crashes their car into a tree or lamppost knows, such heavy objects do not move. How do the bees survive and continue flying?

* * *

When Stacey Combes recruited her postdoc Andrew Mountcastle to study insect flight, she was inspired by a study she had read. In 2011, University of Calgary biologists Ralph Cartar and Danusha Foster had placed video cameras and microphones into fields of flowers that were visited by bumblebees. Slowing down the recorded sound, they occasionally heard a strange rattling sound that came in pulses. A similar sound can be heard if you stick an object into an electric fan. What could this sound be? When they paired the pulses with videography, they concluded that it was the sound of a bee's wing striking a flower 50–100 times per minute. This regular crashing is responsible for the edges of a bee wing growing progressively more jagged over time.

An airplane wing would not survive a single crash that the bumblebees suffered repeatedly. This is in part because of the low weight of the bee's wing. While airplanes generate their lift using propellers, insects flap their wings, and must do so very rapidly to generate forces sufficient to support their weight. Because a bee has to move its wings two hundred times a second, its wings are as light as possible. In fact, they can weigh somewhere between 0.5 to 6 percent of the insect's body weight.

A lifestyle of constantly hitting obstacles is difficult to imagine. But it makes more sense if you think of life from the perspective of a flying insect. A bee cannot afford to slow down and avoid crashes, due to energetic reasons. As it flies, its energy sources are constantly being drained. It uses 10 times more energy flying than walking. Even if it were to stop in midair, it would have to hover, using energy at an even faster rate. Moreover, it has the energetic demands of its colony to deal with. To keep the brood in its colony alive and growing, a honeybee must return from each trip with up to 30 percent of its body weight in pollen. With this grueling schedule and the high energetic costs of flying, a bee does not have time to stop and think. It must get from A to B as fast as possible, and no matter what obstacles lie in the way.

To understand how insects can strike plants without breaking their wings, Stacey and her postdoc Andrew had to design controlled crash tests for insects. They first caught bumblebees and yellow jacket wasps and anesthetized them using cold air. They built a brass straitjacket for the insects. The straitjacket kept the insect's wings in the open position and protected it from its own struggling. Now that the insect was staying still, they could gently apply force to the wing, simulating an impact. When you push on an insect wing, the creases in the wing causes it to bend in predictable ways, as if it were an origami crane.

An insect wing comprises two membrane layers like the bread of a sandwich, held together by a two-dimensional network of veins. The vein network makes the wing like a paper crane that has been unfolded and spread flat, and each insect species has its own particular pattern of veins. Both bumblebees and yellow jacket wasps have veins and a *crumple zone*, a clear crease formed in the wing when its deformed. When an insect wing strikes a branch, the crumple zone bends reversibly as the wing makes contact. Then when the insect pushes away from the branch, the stored energy in the wing's veins causes the wing to reopen, assuming a flat shape. Figure 6.2 shows the vein pattern of

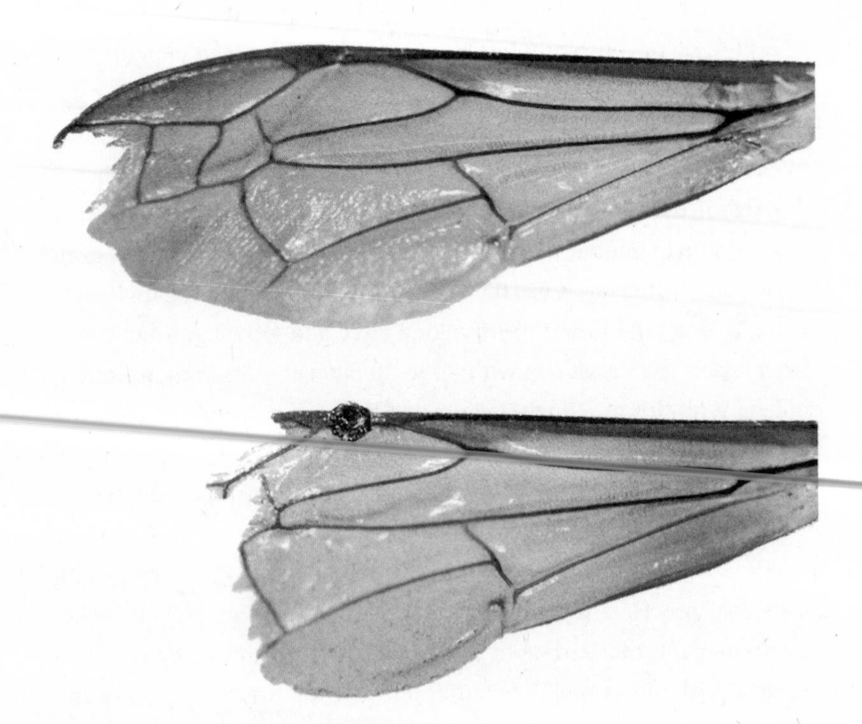

FIGURE 6.2. Above, a yellow jacket wing, and the network of veins that form the wing's crumple zone. The veins are made visible by their embedded resilin, a rubbery material that fluoresces with ultraviolet light. Below, a hexagonal piece of glitter is used to splint the wing. After over 300,000 collisions, an unsplinted wing remains mostly intact while a splinted wing is damaged. Courtesy of Andrew Mountcastle.

a wasp, given by the dark lines, and a light diagonal crease where the wing crumples upon impact.

Stacey and Andrew designed a splint for the bumblebee's wing, and applied it right to the crumple zone. The splint would effectively stiffen the wing and prevent the crumple zone from working. They also wanted to test the bee's flight with the splinted wing. Therefore, the procedure could not substantially increase the weight of the wing.

A bumblebee's wing weighs about as much as a sesame seed, and the splint had to weigh substantially less than that. Since the wing was being flapped back and forth, any additional weight to the wing would alter the bee's motion.

Andrew tried many options, including a wooden splint and dabs of glue, but they were either too weak to restrain the wing or too heavy to allow flapping. A summer student in the lab saw him toiling under the microscope in frustration. She asked him, did you try glitter? Under the microscope, Andrew saw that plastic glitter was ultrathin, and each piece weighed only a hundredth of a sesame seed. The cheapest color of glitter on the internet was ultramarine. He bought a single vial of extra-fine polyester ultramarine glitter, more than enough for his experiment.

The glitter was only half a millimeter in diameter. Trying to grab it with tweezers was impossible, as the tweezer tip was too large. Material scientists had previously dealt with fragile glasses or ceramics that could not be picked up easily. To that end, they developed Crystalbond, a reversible adhesive that flows when heated and cures when cooled. Andrew applied Crystalbond to the tip of tweezers; heating the tweezers triggered release of the glitter.

Once the glue on the wing had dried, they released the bee and tied a string of plastic beads to its body. The beads trailed down from below the flying bee. By counting the number of beads that were off the ground, Andrew could measure the maximum amount of load the bee could carry, and in turn the effect of wing flexibility on this carrying capacity. This was an important experiment because at the time, there were a number of theoretical studies trying to understand if wing flexibility was beneficial or not for the insect. While previous researchers had used computers to study the forces on flexible wings, incorporating the entire complex wing stroke was not easy. Andrew's experiment could make clear once and for all the effect of wing flexibility.

To Andrew and Stacey's surprise, the flexibility of the wing had a measurable effect on the bee's carrying capacity. A bee with a flexible wing could carry 10 percent more weight than one with a stiff wing. This meant that simply by having crumple zones in the wing, the bee could use 10 percent less energy.

Andrew and Stacey were convinced too that the wing's flexibility had a role in the insect's surviving impact. They applied glitter splints to the wings of both bumblebees and wasps, and found that the splints had a larger effect on stiffening the wasp's wing. They then proceeded with impact tests on wasps. They kept the wasp in the same straitjacket that they had built and then spun it like a merry-go-round. Every time the wasp spun, it collided with an object. This simulated the impacts the wasp experienced over a lifetime. After 30 minutes of spinning, or 400,000 collisions, they could observe obvious wear in the wing, as if someone had taken a bite mark out of it (Fig. 6.2). If they splinted the wing to make it stiffer, the bite mark region was much larger, indicating that the flexibility helped the wing to withstand impacts. The crumple zone thus acts like a shock absorber, permitting the wing to bend without breaking. Designing wings of flying robots with crumple zones might help them fly farther with less energy and crash into objects without damage.

How can an insect wing survive 400,000 collisions? One reason is because of the special material in the crumple zones, patches of something called resilin. If you shine ultraviolet light on a bee's wing, the patches of resilin shine. Resilin is a super-springy material. It can be stretched up to 300 percent its original length, and returns up to 97 percent of its original energy. It is the closest thing on earth to a perfect spring. The closest material we can manufacture is Zectron, the primary component in the Super Ball, which can bounce up to 81 percent of the height it was dropped from. In comparison, a golf ball bounces up to 74 percent of its original height, a tennis ball 50 percent, and a wooden ball only 35 percent. A synthetic bio-compatible resilin would be a game-changer for sports. Vertebrates rely on the much inferior tendon, which returns only 90 percent of stored energy. If an insect's resilin replaced our tendon, we would travel 10 percent farther with every step we took.

Some insects have gone beyond just a crumple zone in the wing; they have made their entire body into a crumple zone. It enables the insect not only to escape brief impacts, but to completely morph its body into different shapes, according to its environment. While such an idea sounds like science fiction, such insects are around us, even when we think we are alone. They are the secret residents of our houses, the raiders of our kitchens. And they use this ability to sneak into our houses, no matter how small the crevices are.

* * *

It was dusk, and Berkeley biology student Kaushik Jarayam heard cockroaches skittering. He was eating a bagel as he did every morning, sitting on the steps of the Valley Life Sciences Building. Crumbs from Kaushik's bagel were beginning to litter the concrete steps. The skittering grew louder and turned into rustling as the leaf litter from a nearby dogwood tree began to move. He saw two long antennae poke out of the undergrowth and then a flash of brown ran toward his leg. Reflexively, Kaushik stood up and did a little dance, stomping his feet. Most of his strikes simply hit concrete. But one foot landed on a small creature, emitting a squishing sound. The creature ran away, darting into one of the small cracks in the stairs he had been sitting on. It was *Periplaneta americana*, the American cockroach, and it was seemingly invincible. A shoe had smashed it into the ground, yet it could still run away. Kaushik was intrigued. How could a cockroach survive a pounding like that? By performing more careful experiments, Kaushik would learn that these animals, which appear hard, are actually soft. On the way he would build a crushable robot, one designed to bend but not break.

Kaushik was studying for his PhD in biology, although he had never taken a biology class. He grew up in Bangalore, India, which had been a small town when he was younger, but had since turned into a bustling city, flooded by IT professionals. He had attended the Indian Institute of Technology Bombay, known as IIT B, and focused on mechanical engineering, in particular manufacturing. In our world, evidence of manufacturing is everywhere. Plastics are poured and molded and metals are cast, most of them designed to be hard. As Kaushik was beginning to see, the frontier of manufacturing was not in making machines and devices harder, but in making them softer.

Kaushik's interests in manufacturing coincided with the development of the quadrotor, helicopters with four independent blades, arranged as quadrants of a square. The number of blades make the quadrotor highly maneuverable. The quadrotor can spin in place, swoop, dive, and hover, starting and stopping on a dime. The problem with early quadrotors was that their power far exceeded their survivability. They were so fast that if they ran into things, they would simple shatter into pieces. Engineers began to envelop the quadrotors in a kind of flying hamster wheel that would crush and protect the blades if a crash occurred.

Kaushik started studying the motion of cockroaches, which seemed indestructible with regard to crashes. Such an ability is necessary, from a cockroach's point of view. A cockroach is in constant danger of being eaten by predators. Lizards, cats, birds—all these animals would love to feast on the large source of protein and fat that is the cockroach. Once a cockroach is caught, it is quickly chewed up and swallowed. The cockroach's only chance for survival is to accelerate and run as fast as possible.

A cockroach is always ready to run for its life. It can react in 1/50 of a second, ten times as fast as a human. It can run at 25 body lengths per second, or the equivalent of a car traveling at 280 miles per hour. Speed is the key to its survival, and is so important that the cockroach has little time to avoid objects. It simply runs head first into them. If you slow down the action with a high-speed camera, you will see a cockroach run head-first into a wall, bounce off it, and then climb straight up the wall. If a small crack is below the wall, it wedges its body into the tight crevice as quickly as possible. This behavior gets it out of sight of its much larger predator. Often, it just barely succeeds. Tara Maginnis, a biologist in Montana, once took a census of insects found in the wild. Insects are born with six legs. Among batches of wild-caught insects, the average number of legs is five. These five-legged insects are the lucky ones, barely escaping to skitter another day.

Since cockroaches were too fast to study in the wild, Kaushik constructed an obstacle course in the laboratory. A cockroach's preferred home is a forest floor littered with leaves and debris. In such surroundings, the brown and black coloration of its body camouflages it perfectly. Kaushik built an open trackway that ended at the entrance of a tunnel. He made the roof of the tunnel only the height of a stack of two pennies, which was one-fourth of the cockroach's standing height. A cockroach entering the tunnel would be like a golden retriever trying to cram its body into a mailbox.

Using a high-speed video camera, Kaushik filmed the cockroach entering the tunnel, which from afar looked like a small crevice. The cockroach stuck its long antennae in first. After exploring the space, it paused briefly. Then it ducked its head into the crevice to take a look. Sometimes it rammed its head multiple times to get it in. The cockroach was 1.2 centimeters tall, four times as tall as the crevice. To force its way in, the cockroach walked forward with its front legs. The crevice was so low that as it put its head into it, its body tilted upward 45 degrees, causing its back legs to flail in the air. Unperturbed, it continued pulling its body in with its front legs. In a single second, the entire cockroach was inside the crevice. From the predator's point of it view, the cockroach might as well have disappeared.

Kaushik lined the walls of the tunnel with glass, so he could see inside. The cockroach had squeezed itself to nearly nothing. Its legs, which are usually arranged underneath its body during a standing posture, were now sprawled horizontally like a crab's. He used a stick to poke inside the entrance of the tunnel, imitating a cat sticking its paw into a hole. Surprisingly, the cockroach did a crab walk away from the stick. With more poking, the cockroach picked up its speed and began to run, although its body was nearly squished flat.

The ability of a squished animal to run at full speed was a surprise to Kaushik. When we drive a car through a narrow alleyway, we don't drive at full speed. At full speed, the car could easily be damaged beyond repair. While the cockroach appears to be covered in shiny armor, it has a number of soft joints that permit its body to be surprisingly flexible. For instance, each joint of its legs is composed of a flexible semi-transparent membrane, just like the overlapping lamellae on the knees and shoulders of a medieval knight's body armor. The belly of the cockroach is also covered in overlapping plates like venetian blinds. Kaushik thought that these plates might allow the cockroach's body to expand laterally when it is squished. The cockroach was born to be squished.

To test its limits, he put a cockroach in a mechanical press with transparent walls on all sides to prevent it from escaping. He then applied a press up to 900 times the roach's body weight, the equivalent of crushing a person beneath a one-bedroom apartment. The belly plates gently expanded, allowing the soft innards of the cockroach to push outward through soft transparent membranes. Indeed, the inside of the cockroach was mostly fluid, and so downward compression meant that the fluid had to push out somewhere, and it did so between the plates. When the force was removed, the transparent membranes pushed the cockroach back to its normal shape. The cockroach was able to walk away unharmed after such a load. The cockroach was like a motorized water stress ball, able to keep walking, even when it was squashed unrecognizably flat.

The maximum survivable downward push on a cockroach is limited by how much its exoskeleton can withstand the pressure of its fluid contents. However, robots have no such constraint. They can be made of interconnecting parts with air spaces inside rather than liquid. Just like an origami crane, a robot has the potential to squish down to a single sheet of paper and still function. Inspired by the cockroach's crushability, Kaushik decided to build a synthetic version.

The years before had seen a revolution in six-legged or hexapedal robots. There had always been interest in mobile-legged robots, but interest came to a peak when DARPA, the Defense Advanced Research Projects Agency, gathered several investigators for a special meeting in 2000. DARPA was known among government agencies for funding moonshots, ambitious projects that would push the boundaries of the field. That year, there was an interest in legged robots with the mobility of insects. Kaushik's advisor Bob Full was there, as well as Dan Koditchek, an electrical engineer from Michigan. Dan watched Bob's videos of robots traversing rough terrain and was inspired to build RHex, a six-legged robot about the size and weight of a 15-pound bulldog. It traveled over pebbles, grass, and other obstacles, all done so in open loop—without feedback from the environment. It could run blind without stumbling.

In 2009, the microfabrication industry met robotics research to build a series of ultralight hexapedal robots. One of them, called DASH, or Dynamic Autonomous Sprawled Hexapod, was cut out of a sheet of card stock, making it only 30 grams, easily sitting in the palm of your hand. It was designed by Paul Birkmeyer, a student of Ron Fearing, an electrical engineering professor at UC Berkeley. Birkmeyer and Fearing used a method called the *smart composite microstructures* (SCM) *manufacturing* technique, where stiff parts could be linked with soft parts to build a composite robot. First, a computer was used to sketch out a blueprint, an arrangement of cuts on a planar sheet. A laser cut makes these cuts onto a piece of poster board that is then folded in half, to sandwich a piece of flexible polyester sheet in between. Adhesive and heating is applied to make the sandwich permanently bonded together. A laser cutter then cuts out holes in the folded sandwich, yielding flat shapes that can be bent and folded like a pop-up book. The final three-dimensional shape had six legs that could be actuated together by a single small DC motor similar to the kind found in toy remote control cars. It could travel a body length per second, or the equivalent of a car at 10 mph, and its soft exterior made it perfect for redesigning into a crushable robot.

DASH had a fundamental problem that stopped it from being crushable. It could only be squished down to the height of its motor, which was necessarily hard. Motors made completely of soft rubber had not been developed at that time. Kaushik's innovation was to drive the robot with two separate smaller motors on the left and right side of the robot. Each motor would drive the three legs on its side. Since each motor only had to drive three legs instead of six, it was smaller than the original motor.

The original DASH had a rectangular base where its legs were hinged. Kaushik designed a fracture zone in the middle of the robot, enabling the robot to press down yet spring back up (Fig. 6.3). Imagine two bases that are connected by a spring in the middle. Pushing down with your finger allows the two halves of the robot to splay outward. When you stop pushing, the robot pops back up. To finish it off, Kaushik covered the top of the robot with a polyester sheet folded into a compressible shell, like an origami hat. He lubricated this shell with oil so it would provide low friction to the ceiling of the tunnel.

When the robot was freestanding, it walked easily on solid ground. The entire robot was palm-sized, weighing only 50 grams, the weight of a few grapes. It was autonomous, with its own battery and onboard electronics. Its body was made from laminated paper so it could be picked up, dropped, and even bent and twisted by hand. Kaushik then challenged it by putting it into a tunnel half its height. As he designed, the middle of the robot's back bent, allowing it to flatten. However, the recoil of the spring in the robot pushed too hard against the ceiling and the floor, creating large friction forces that pinned the robot in place. Its small legs scratched uselessly against the ground as the robot attempted to push itself forward.

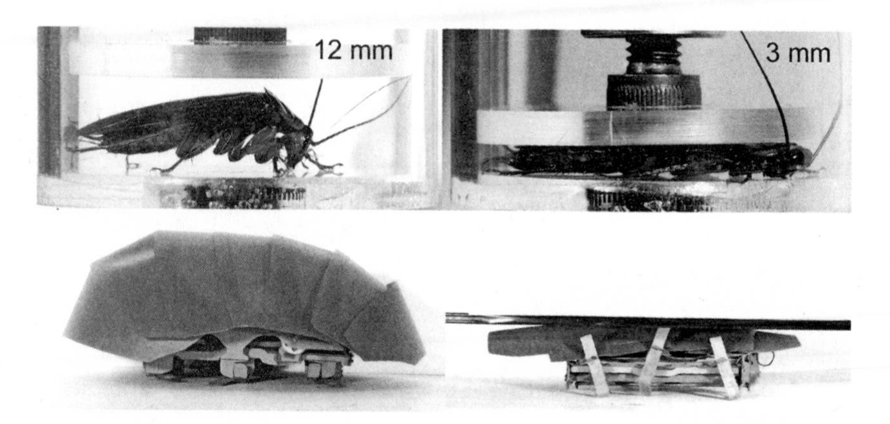

FIGURE 6.3. The bio-inspired robot CRAM, or compressible robot with articulated mechanisms, along with its natural counterpart, the common cockroach *Periplaneta americana*. The robot can be squished half its standing height and still propel itself, while the roach can be squished to one quarter its height. Courtesy of Kaushik Jarayam.

The split motor and deformable spine had worked well, but now the legs were the problem. He needed a redesign of the legs. His current design involved paper legs like matchsticks that the robot would walk on. But when the robot was squished, these matchsticks bent away from the body, preventing the feet from getting sufficient grip. When the robot was squished, the extra friction against the top and bottom walls resisted forward motion. The legs were now in an awkward position, and moreover, the ceiling pushing down on the robot meant that even more force was required from the legs. Kaushik went back to watch his videos of the cockroach crawling.

The cockroach's legs, like its abdomen, were collapsible. While free-running, the tips of the feet touched the ground. In tunnels, however, the legs sprawled outward as if the cockroach was doing the splits. Its knees pushed off the ground as we do in a crawl. As soon as the cockroach exited the tunnel, elastic energy stored in the legs pushed the cockroach back to its feet.

Kaushik realized that the legs of his robot also had to have a collapsible design so that they could grip the ground both in the freestanding and in the squished posture. He designed L-shaped legs by bending matchsticks in half. He then made the hip connection with the leg even more flexible. In the freestanding position, the robot walked on one prong of the L-shaped leg. When it was squished, the legs splayed out, allowing the robot to walk on the other prong of the L. This design enabled the robot to grip the ground, no matter what configuration its legs were in.

Kaushik's collapsible robot may have applications in search-and-rescue operations. After an earthquake, first responders want to evaluate if rubble contains survivors, but the problem is that this rubble is often unstable and may be too dangerous for people to walk upon. Small collapsible robots like Kaushik's could be released in large numbers to scurry through nooks and crannies, searching for survivors with sensors. Kaushik's robot is mostly made of cheap materials, paperboard and motors for toys. This means the search-and-rescue robots could also be expendable, without the need to retrieve them after they accomplished their tasks.

* * *

All the animals and robots in this chapter showed an ability to deal with impact with obstacles. They relied upon their physical design, the ability to bounce off or collapse without being injured. In the next chapter, we look at how neural systems work, and how an animal must think in order to move well.

CHAPTER 7

The Brain behind the Brawn

To demonstrate fruit fly aerodynamics, Cornell University physicist Jane Wang drops sheets of paper from the top floor at McGraw Tower. The sheets fall slowly at first but pick up speed and begin bending, tumbling, and fluttering. The chaotic motion is driven by the trail of invisible eddies created by the paper. The paper is so lightweight that these eddies can in turn influence the path of the paper. This feedback leads to the paper falling more and more unpredictably. Eddies like these are everywhere. Generated by air flow past tree branches and other obstacles, such eddies are usually too small for us to notice. But for an insect like a fruit fly, they come in like the linebackers of a football game.

This chapter is about how animals automatically respond to their surroundings. When you drive a car, you use a similar system known as cruise control, which allows you to maintain a constant speed on the highway. Imagine if your car was inherently unstable like a kite. If you let it coast down the road, its low weight would cause it to rattle back and forth, swerving along the highway. This is a daily reality for the fruit fly. For it to fly at all, it must have constant feedback between its sensors and its wings.

In 1996, Jane Wang finished her graduate study in physics at University of Chicago. She was a newly trained expert in turbulence, a regime of fluid motion where flows can be fast-changing and chaotic. She continued her work on turbulence as a postdoc at Oxford University. One day, as she was reading a paper at the Mathematics Institute Library, she decided to take a break. She took a stroll around the fluid mechanics section and a thin book with an intriguing title, *The Mechanics of Swimming and Flying*, caught her eye. She was amused, and kept the book in her mind as she visited New York University, the home of the author Steve Childress, whose office was filled with paintings of fluid dynamics and their governing equations. In his soft-spoken Texas twang, he said, insect flight is a very confusing field. These words stuck in Jane's mind, and for the next year she stayed as a visitor at NYU, applying her knowledge of turbulence to try to understand just how insects can fly (Fig. 7.1).

The big difficulty with modeling insect flight is that an insect's wings are paper-thin, making it especially difficult to resolve the flows at the wing's trailing edge. But it is at this sharp edge where the magic of insect flight happens. The boundary layer around the wing sheds a vortex, leading to enhanced lift. The problem is that flows around sharp tips, also known as *singularities*, are intrinsically difficult to solve. The equations that govern fluid flow, known as the *Navier-Stokes equations*, have yet to be solved on paper. In fact, the Clay Mathematics Institute offers one million dollars for a proof showing existence and uniqueness for these equations. Jane chose to use a computer to approximate solutions to the Navier-Stokes equations in order to predict the fluid motion around the wing tip. Because the tip is so sharp, the air space around it must be broken up into a great number of very small boxes in order to model the fluid motion accurately. But employing a large number of boxes would be challenging for the computers of the day.

Jane's strategy to resolve the motion at the tip was to study the wing in two dimensions. If her computer could only resolve a million points,

FIGURE 7.1. Multiple exposures of a fruit fly turning to escape an encroaching object, with approximately five wingbeats between each image. The fruit fly changes its direction within a few wing strokes by making small adjustments to its wingbeat pattern. Its body is inherently unstable, and it can only maintain stable flight by sensing and correction of flight errors with changes in each wingbeat. Courtesy of Florian Muijres.

she could do so with 1000 x 1000–point accuracy in two dimensions, more than 10 times the resolution of three dimensions, where computer resolution would limit her to a box 100 x 100 x 100 points in size. Studying things in two and three dimensions would always lead to two camps. The two-dimension camp believed in insight and intuition based on ideals, the three-dimensioners believed in absolute accuracy. Before Christmas 1998, she submitted her computer code to be solved by the NYU supercomputer NESCE. When it returned, she was happy to see that her computer code had resulted in a vortex being shed from the wing, a hallmark of the experiments of insect flight.

Over the next ten years, Jane would use her computer code to solve for a variety of complex flows related to insect flight. Insects move their wings through the air in a figure-eight motion. The flows generated are similar to those behind playing cards fluttering and tumbling through the air. During this period, she recruited experimentalist colleague Itai Cohen and his graduate student Leif Ristroph, both of whom expressed interest in insect flight. Together they devoted their combined talents to investigating the free flight of insects, as distinguished from studies of insects tethered to a wire, the dominant style of study of the day. Leif was a tall Texan who was a whiz in the lab. He carried out the experiments, and Jane and her group designed tracking software to extract the wing and body motions, as well as making analyses of the experiments.

Insect flight has long been plagued by the difficulty in tracking fast-moving insects. Insect wings beat hundreds of times per second as they zip across a room. A camera may be able to catch the trajectory of the insect, but not the insect's wingbeats. To solve this problem, Leif created a sports stadium for fruit flies using a Plexiglas chamber about a foot in length, in which he put nearly a hundred fruit flies. When the flies flew through the box, free flight was recorded simultaneously by three high-speed cameras. The cameras had enough resolution to see the position and angle of the wings and body of a fly. Jane thought

Leif's experiments were a good way to answer questions about insect stability and maneuverability, which was a poorly understood aspect of insect flight.

When we look at an airplane, the most important apparatuses for flight appear to be the wings. But if that were the case, flight would have been invented far earlier than 1904. The Wright Brothers' big contribution to flight was not wing design, which was well understood. It was the control of the airplane. They accomplished this using an airplane with flexible wings and numerous independent flaps, rudders, and ailerons, which they practiced moving simultaneously. As the Wrights flew the plane, they constantly tweaked each of these flaps to keep the plane in the air, making the plane difficult to fly by anyone else but them. Over the next ten years, prolonged flight was only made possible by automatic stabilizers involving gyroscopes that automatically sensed a plane's heading and adjusted its flaps.

Behind the wings of fruit flies are a pair of miniature former wings called *halteres*. As the fruit fly flies through the air, it flaps both its wings and its halteres in unison. While the wings propel the fruit fly, the halteres act like gyroscopes, enabling the fly to sense its rotation. Even if the insect rotates its body, its halteres have inertia, or a memory of what direction they were originally traveling in. They tend to continue vibrating in the same plane even if their support rotates. Moreover, the bases of the halteres have fine sensors that can measure the perceived forces on the halteres as the insect rotates. These sensors are so fine that as long as the halteres continue flapping, the insect can measure the rate of change of rotation in three directions, pitch, roll, and yaw.

To test how well these halteres worked, Leif glued one-millimeter pieces of wire to the back of a fly and used magnets to trigger these wires to provide a brief 30-degree horizontal rotation to the fly. He and his team named this rotation an aerial "stumble." It was as if the fly had indeed misstepped and was pushed briefly off its trajectory. They found that the fly was able to correct and return to its original trajectory within three wingbeats. The bigger the stumble, the more time it took to recover. The wing and body angles need to be tracked precisely for analyses of flight. Jane's student Gordon Berman adapted an algorithm in computer vision to track the wing and body motions automatically. Jane and her student Attila Bergou then used Jane's expertise in flight aerodynamics to construct a model of a fruit fly on computer. She incorporated the wing motions that Leif measured.

Jane further developed her computer model to simulate flight in six degrees of freedom—basically, all the ways a fly can rotate and translate in space. She used her computer model to test different feedback control loops to mimic an insect's neural feedback circuitry. She was particularly interested in the insect's reaction times, the time scales in the feedback circuitry. Strangely, the computer model showed that the fruit fly would have difficulty flying as well in quiescent air. She watched as a computer fly tried to hover in midair. It paddled its wings back and forth like a rowboat, getting lift on both the upstroke and downstroke. However, as it paddled, its body began to pitch back and forth like a teeter toy. Because the wings were attached to its body, this motion of the body fed back to the wings, changing the angle that they stroked through the air, and thus the direction of the lift generation. The fruit fly began to slowly arc downward and lose control, much like a piece of tumbling paper. Her simulations showed that a fruit fly needed to have feedback just as much as an airplane needed a pilot.

Jane's simulations showed that the feedback provided by the halteres was absolutely necessary for flight. The problem is that halteres still need time to work: after information is collected by the fly's halteres, the information must travel to the wing muscle, which are of two types. Large muscles flap the wing and smaller steering muscles can quickly make small adjustments when necessary. The fly can only react as quickly as its halteres and eyes can sense and its steering muscles can flex. When Jane and her student ran the simulations, they found that the simulated fruit fly could only maintain flight if it sensed its world frequently enough. If it sensed too infrequently, after a few wingbeats the fly would veer off course and crash. One interval that worked well was the fly sensing once per wingbeat, which is quite a high sensing rate: a wingbeat occurs 200 times per second. Without this beat-to-beat sensing, in a single blink of a human eye a fruit fly would have flapped 60 times, and already have careened out of control.

One might think that flight is inherently tricky because it has so many degrees of freedom. A fruit fly can pitch, roll, or yaw, rotating along any of three axes. Perhaps it is understandable that flight would require immense amounts of feedback from its surroundings. However, animals on the ground are in need of control as well.

* * *

Under the kitchen table, a cockroach was nibbling on a cookie when it detected footsteps. As the footsteps drew closer, it began making its way toward the kitchen wall, relying on its long, thin antennae for guidance in the dark. An antenna grazed on a leg of the kitchen table, which it sidestepped easily. The footsteps grew closer. The cockroach began to panic, doubling its speed to the human equivalent of 100 miles per hour. As it approached the kitchen wall, one of its antennae bent upon contact with the wall, indicating a crash was imminent. The cockroach turned sharply and ran on alongside the wall, continuing at full speed. As it ran, the rough wall dragged the tip of its antenna backward, deforming the antenna into a J-shape. When the cockroach ran into a corner, it veered again, turning at full speed like an Indy car driver.

To understand how a cockroach moves so quickly in the dark, Johns Hopkins University engineer Noah Cowan planned to reverse-engineer the cockroach antennae. Noah is an expert of control systems, and demonstrates his expertise using the lightweight juggling pins that he has juggled since middle school. He can take one and balance it at the tip of his nose indefinitely. The feat is a result of feedback. The bowling pin doesn't stay balanced unless Noah measures the sway of the pin with his eyes, and moves back and forth to balance out the pin's motion. The cockroach employs a similar strategy, using its antenna to measure how far the wall is, and then turning back and forth to stay at a constant distance from the wall.

Noah first started studying cockroaches years ago when he was a postdoc in Bob Full's lab at UC Berkeley. As part of his initiation into the lab, Noah was asked to reach into a tank of dozens of cockroaches and to take one out for his experiments. He gingerly reached his gloved hand into the tank. As his hand touched down, a cockroach darted up his arm and onto his shoulder. Noah screamed, swinging his arm wildly, flinging the cockroach into the air—it started flapping its wings and miraculously floated back down to the ground. After Noah calmed down, he practiced picking up cockroaches, eventually learning to hold onto a cockroach without its escaping.

As he watched cockroaches move about in their tank, he noticed they preferred to follow walls, and when they did, they relied on their antennae (Fig. 7.2). A cockroach's antennae are as long as its entire body. Each antenna has a swivel hinge at its base, which the cockroach uses to swing both antennae like two white canes searching for obstacles. This search behavior changes as soon as the cockroach touches a wall. The once waving antennae turn rigid, and are held out horizontally at 45 degrees to the cockroach's centerline. This antenna posture apparently helps the cockroach to find corners and other changes in the wall's position, but until recently little was understood of how this was accomplished.

To better understand how the cockroach turned corners, Noah worked with his first PhD student, Jusuk Lee, to build a racetrack like the Indy 500. Race cars begin the Indy with what is called "first turn," a straight road leading to a 90-degree turn, with raised curbs on the inner and outer edges to protect the incoming cars. Cars take the turn at over 200 miles per hour without slowing down. In Noah and Jusuk's racetrack, the roaches were flying along at 25 body lengths per second,

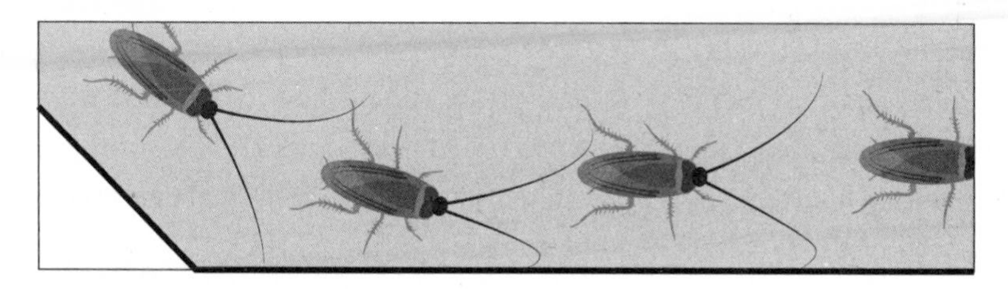

FIGURE 7.2. Multiple exposures of a cockroach running full speed in the dark, with one image for every two strides. By holding its twin antennae ahead of it, the cockroach can measure its distance from the wall using deformations in one antenna. A feedback system senses the deformation and enables the cockroach to avoid both crashing into the wall and veering too far away from it. Redrawn from a figure by Noah Cowan.

or the equivalent of a car traveling at 280 mph. Noah covered the cockroach's eyes with white-out to ensure that it would only use its antennae to navigate. He then released the cockroach, and it took off. An angled wooden wall placed on the outer edge of the turn enabled the cockroach to track the wall with one antenna. He watched as it ran along the wall. Since it could not see the turn, it did not slow down as it approached it.

At the speeds that the cockroach was running, it would have 1/25th of a second to recognize the corner and turn, or otherwise risk crashing into the wall. He filmed the roach attempting sharp turns of 90 degrees, and also gradual turns of 40 and 30 degrees. The 90-degree turn turned out to be too challenging for the cockroach. It simple crashed into or ran straight up the wall. However, more moderate turns of 30 and 40 degrees were easily maneuvered. As soon as one antenna sensed the turn, the cockroach's entire body began to angle appropriately. Noah and Jusuk decided they would focus their efforts on analyzing the cockroach that took moderate turns.

Jusuk used a refrigerated plate to chill the cockroach down, which put it to sleep. Then he painted two dots on the cockroach's back, one at the head, the other at the tail. By tracking these dots, they could calculate the angle of the cockroach's body and its distance from the wall. From these data, they were able to estimate how quickly the cockroach detected the turn, and, more importantly, the subtle ways the cockroach made its correction—the dynamics of the correction.

The cockroach's neural feedback is like a car's cruise control, a sophisticated form of feedback system. Such systems do more than just produce an output at a given stimulus, such as the knee-jerk reflex. Instead, they can target a state in the body and make it stable. Driving down a highway, cruise control measures a car's speed and the rate that speed is changing. It uses both these data to bring the speed back to the desired state. In a similar way, the cockroach measures both its distance from the wall and the rate at which this distance changes. The cockroach needs both of these measurements to keep stable as it runs. Distance from the wall is used in *proportional control*, in which the cockroach adjusts its speed proportionally to the distance it measures from the wall. The rate of change of distance is used in *derivative control*, adjusting its speed relative to the rate of change of the distance to the wall—the derivative. Why are both proportional and derivative control necessary?

Think of when you are driving your car up a driveway to get it into your garage Let's say you push down on the gas pedal in proportion to the distance you are from the garage. Doing so will certainly allow you to reach the garage. But at the instant you get there, you will also have accumulated speed, and you might crash into the garage. This shows that you can't just look at your position to control the gas; you must also keep track of your speed. That requires a derivative controller. The derivative controller tells you if you're approaching the garage too quickly. If so, the controller tells the car to start slowing down. The proportional and derivative controller work in concert so that the car stops right at the garage. In the same way, the cockroach hits a turn and is able to maintain its distance from the wall without losing contact with it.

Now that Noah was confident he understood how the neural feedback worked, he began working on the physical interaction between the antenna and the wall. For most robots, this second part is ignored. This is because most robots are programmed to specifically avoid interacting with obstacles. They roll on rubber wheels atop linoleum floors using sensors to avoid contact with anything around them. Some have primitive "bump" sensors that tell the robot to stop and change direction in a crude manner compared to a cockroach. However, in nature, there are often a large number of obstacles around animals as they traverse the wild. To truly reverse-engineer antennae, Noah and his team would have to understand how the antenna interacted with the wall physically.

Noah began with a key observation made by his student. Jusuk noticed that cockroaches cannot follow all types of walls. When they were conducting his experiments, they first tried using slippery Plexiglas walls. When the cockroach pressed its antennae against the wall, its interaction with the wall was unstable. Its antennae easily flipped from the intended J-shape to a straight orientation, which rendered the cockroach blind to its touch sense. Without the bent configuration, the touch sensors cannot measure an antenna's position. The cockroach quickly either slams into the wall with its body or loses touch with the wall altogether, frantically running in random patterns to find it again. As a result, cockroaches were clumsy when tracking Plexiglas walls, finding the wall only by crashing into it. On the other hand, the cockroach easily followed rough fiberboard, an analogue to the rough obstacles like tree bark that it would find in nature. Along fiberboard, its antenna quickly bent from its straight shape to a J-shape. Because the antenna's barbs gripped the wall as the cockroach ran forward, Noah had a hunch that the roughness of the wall was important in getting this shape.

To probe this idea further, Noah asked his previous postdoc advisor, Bob Full, if his lab could "shave" the tiny hairlike spines off of a cockroach antenna. Bob's PhD student, Jean Mongeau, took on the challenge. The antenna spines are like a cat's tongue, which feels smooth when rubbed back toward the mouth, but rough when rubbed toward the tip of the tongue. The spines on the antenna act like small hooks, and Noah had a hunch that these spines were responsible for the antenna hooking so easily onto walls. Jean successfully used a laser to shave the antenna, leaving the center stalk intact. Without these hairs, the antenna slipped on any surface it touched, unable to assume a J-shape and effectively blinding the cockroach. Shaving the antenna was like replacing a car's rubber tires with smooth glass tires with no grip.

Inspired by these new findings on the antennae, Noah's team, headed by his PhD student Alican Demir, began working on a wheeled robot equipped with antenna feelers. The robot was envisioned for use in emergency disaster situations to make its ways around buildings whose electrical lighting systems had been shut down. The base of the robot was a three-wheeled tractor-looking machine that was commercially available for those who wanted to test new algorithms for robotic control. Noah's first version of the flexible antenna had four sensors placed along its length to sense bending. The antenna was cast from flexible urethane plastic. They were surprised how poorly the antenna worked. If the antenna was too stiff, it would not bend when it touched the wall. If too soft, the antenna bent behind the robot, trailing like a blanket, useless for probing the way ahead.

Noah went back to look at the cockroach's antenna. With just a touch of his fingertips, he bent it back. The antenna's long length allowed it to bend with just the smallest force. This length was not feasible in a robot because of the cost of each of the sensors. But a feature that could be imitated was the antenna's shape. Instead of having a uniform thickness like a cylinder, the antenna was shaped like a very narrow cone, with a thickness that decreased toward its tip. The wide base of the antenna kept it stiff there, while the thin tip of the antenna helped to accommodate bending. As a result, when the conical antenna was deformed, it bent near the tip. A cylindrical antenna would bend much closer to its base. Bending at the tip is helpful because it allows the antenna to reach farther ahead of the roach. Just as if it were using headlights, the cockroach could sense the area ahead of it with enough time to adjust its motion. Using a new mold, they cast a similarly tapered antenna, which greatly improved the robot's performance. The new antenna was so effective that they could even reduce the number of sensors while keeping the same performance.

Since this work, a number of other investigators have continued to work on using long filaments to sense environments, known as the *active tactile approach.* Much work remains to be done on understanding the feedback loops in the whiskers of other animals such as rats, or seals, whose whiskers are used underwater. The use of such long filaments seems to be a common method among animals. Their use across a range of environments is also inspiring the next generation of touch sensors to help improve the navigation of robots in cluttered terrain.

Sensing is an important part of animal locomotion because it helps the animal deal with an unpredictable terrain. Another equally important part of locomotion is the maintenance of pattern, of the repeated footsteps or tailbeats. This is done by the *central pattern generator,* which is seated in the spinal cord. To understand how the spinal cord works, we will have to turn back millions of years to one of the most primitive animals, the lamprey.

* * *

Dutch robotics Professor Auke Ijspeert sat in front of a steaming plate of stewed lamprey. Each filet was similar in size to filet of eel. It had been blanched to have its slime removed, then stewed for two days and four hours in its own blood. Now it was accompanied by Bordeaux wine and candlelight. Auke wore black and red monastic robes, and sat at a table with twenty other robed, faceless members. Thus began the initiation rites of the Brotherhood of the Lamprey, as it had been done for years in Saint Terre, France. Biologists had told Auke that if he were to truly understand the lamprey, he would have to know the lamprey inside and out.

Auke had thought about lamprey for nearly a decade, ever since he did his PhD thesis on simulations of lamprey at University of Edinburgh. For his thesis, Auke used computer simulations to understand how the lamprey moved with such a little brain. The lamprey is a living vampire that has roamed the seas, virtually unchanged, for 550 million years. An adult ocean lamprey can grow up to 47 inches in length and weigh 2–3 pounds. Its body is long, black, and slimy. One end is a wide, flared tail to aid in swimming, the other end looks like it has been decapitated. Where its head should be is a reddish sucker filled with concentric circles of small white teeth used to attach to sea creatures and suck their blood.

Auke contemplated the lamprey's way of life as he slowly bit into it. It had an earthy taste, not quite like meat or fish, yet with a texture like scallops. Because the lamprey preceded modern fish, it was considered primitive, and for scientists who sought simplicity, this was a good thing. The lamprey was one of the model organisms for neurobiology, the study of how the brain makes the body move. The lamprey was the link between the hidden world of the mind and the visible world of bodily motion. In this book thus far, we have described how an animal interacts with the world outside of its body. But what does an animal think in order to move?

When Auke was a PhD student, he was drawn to the lamprey because of a 1996 article he had read in *Scientific American* about the lamprey neural network. The article was written by Karolinska Institute's neurophysiologist Sten Grillner, another member of the Lamprey Brotherhood, and a professor who since the 1970s had drawn worldwide attention to the lamprey as a model organism for neuroscience. Sten considered the lamprey as the Ford Model-T for understanding how the brain coordinated movement. While the human brain could be thought of as a Ferrari, the lamprey had all the basic parts that a human brain did, and was easier to analyze because it had much fewer nerve cells of each type. Auke was further attracted to Sten's work because of Sten's collaboration with computer scientists to write programs that simulated the lamprey brain. Auke read Sten's papers closely, and the two met each other at conferences. While Auke performed simulations in Edinburgh, Sten worked on experiments on lamprey in Stockholm.

Sten was one of the few neurophysiologists that could perform spinal cord measurements in a living lamprey. The first step in this process was purchasing a lamprey from a local hatchery. The lamprey arrived in a Styrofoam box. It thrashed wildly, attaching to the edge of the box to try to suck its blood. Sten added a teaspoon of the white powder MS-222 to the cooler, and in minutes, the lamprey was under deep anesthesia. He kept the whole animal on ice to allow its body to survive the procedures he was about to perform. He made an incision along the body of the lamprey in order to access the nearly transparent spinal cord, which was only one third of a millimeter deep and more than a millimeter wide. He could then apply electrodes along the exposed spinal cord and record the activity of its neurons. In the isolated spinal cord he could induce activity in the network of nerve cells that normally would generate the locomotor movements, which allowed him to deduce how the intricate network worked.

While Sten was in Stockholm preparing the lamprey, Auke was at his computer in Edinburgh, working on a computer model of the lamprey spinal cord. His computer model was composed, like the lamprey's spinal cord, of thousands of neuron cells that on their own clearly did not embody intelligence. However, when the cells were put together, intelligence emerged. He knew the live lamprey's cells coordinated together to create a *traveling wave*, seen as the wiggling motion when the lamprey swam. The way these cells did this was by falling into groups called *central pattern generators* or CPGs. Each

CPG is like a small clock that generates ticking signals to govern the pace of different activities. The clocks' interactions with each other generate waves of activity that enable our legs to walk, our arms to swing, and our lungs to breathe all at the same time, without a second thought.

A network of CPGs generates a wave much the way a stadium of football fans makes a wave. The first stadium wave was inadvertently created in 1981 by professional cheerleader Krazy George Henderson. He attempted to have one side of a stadium jump and cheer, and the other side echo this behavior. In other words, he was trying to create an oscillation between two parts of the stadium. When he signaled to have one section jump, he found that the section a little farther away missed the signal, generating a small lag. They jumped a few seconds too late, and the fans next to them mimicked them and they too jumped with a lag. The wave of activity traveled around the entire stadium on its own. This was the first stadium wave recorded, and it would become Henderson's trademark routine. The routine shows that oscillations with small errors naturally result in waves when there are larger numbers of individuals involved.

The simplest central pattern generator is a neural oscillator. Instead of an entire stadium, consider two football fans watching the game on a couch. First, the fan on the right stands up. As she sits down, the fan on the left stands up. By simply watching what the other person does and responding as fast as possible, the two fans propagate the wave, which travels back and forth at a steady pace. The pair of fans is an oscillator, like a blinking light. In the spinal cord, an oscillator is a system of neurons that spontaneously generates a repeating pattern with a set period and amplitude. For the football fans, the oscillator is most regular if the fans are the same size. In that case, the oscillator's period is given by the time it takes one person to stand up and sit down, roughly a few seconds. If the fans are sitting in a particularly deep leather couch, the period may be longer. The amplitude of the signal is the difference between a person's sitting and standing heights.

Auke's computer model was not the first to simulate a lamprey CPG, but it was the first to incorporate the behaviors of real lamprey CPGs. A lamprey's spinal cord consists of a series of 100 CPGs arranged like pairs of passengers sitting in the rows of an airplane. Each pair oscillates, generating its own signal just like the sports fans in the living room. Each pair also receive signals from the passengers sitting in front of and behind them, just like in the wave. From an initial kick start from the brain, the oscillators create a wave that travels down the spine. The wave's speed depends on how quickly the CPGs fire and process the signals from the neighboring CPGs. Auke's plan was to code in the behavior of each CPG, and hope that the traveling wave would naturally arise.

Back in Stockholm, Sten inserted electrodes into different parts of the lamprey spinal cord to listen in on them. He measured periodic bursts of activity traveling down the spinal cord, a phenomenon that neurophysiologists called *fictive swimming,* as if the lamprey were dreaming of swimming. It could not move its body because its muscles were anesthetized. This behavior was spontaneous, and occurred even when the lamprey's brain was removed. Sten measured the extent by which the CPG's behavior was modified by its neighbor's frequency and amplitude.

When Auke put these parameters in his computer program, he found that traveling waves emerged spontaneously in his digital lamprey. This suggested that the lamprey's spinal cord is in fact thinking about swimming all the time. It is only inhibitory signals from the brain that prevents the lamprey from actually swimming. Auke encoded the forces from the fluid surrounding the digital lamprey and found also that the lamprey was able to swim with an amplitude similar to that of the live lamprey.

After Auke graduated with his PhD and conducted postdoctoral studies, he continued his work on lamprey as an assistant professor at EPFL in Switzerland. He became interested in testing his simulations in actual physical models, using robots. He began designing a lamprey-inspired robot, Amphibot. It consisted of ten square modules that together controlled the 100 CPGs found in the lamprey. Each module was encased in a yellow plastic cover, and inside was a small motor for adjusting its angle with respect to the neighboring module. The motor represented the muscles in the lamprey. This much was similar to other robots, but the robot's brain was totally different. Each module contained its own onboard computer that behaved like a CPG, simply generating an oscillatory signal and responding to the modules next door. The parameters in the CPGs were taken from the neural recording experiments from the lamprey. In effect, the robot had a lamprey spinal cord inside of it.

With Amphibot designed in this way, most of its body was out of direct reach of the brain, the leading module in the robot. This separation greatly simplified control of the robot. Keep in mind that most long, snakelike robots are not designed using CPGs. In those cases, the robot's brain would have to control all 10 of the yellow plastic modules. However, with the CPGs, the brain simply had a control stick, just like the stick shift of a car, that it used to control the first module. When it was in low gear, the lamprey swam slowly, leisurely oscillating its tail. As it moved into higher gear, the frequency of waves down the tail increased, and the Amphibot started splashing as it sprinted. The robot acted just like Auke's simulation and Sten's neural recordings from the lamprey. In all these cases, stimulating the brain at a high frequency led to faster swimming.

The use of a robot also led to new discoveries that could not have been made on computer. On computer, it was time-consuming to test varying conditions. With the robot, testing different environments simply amounted to play. Auke started by twisting the robot into various configures like a rumpled shirt or sock. No matter how the body was initially configured, it quickly resumed a traveling wave. For example, if Amphibot was curled up into a spiral to hide behind a rock, it could easily wake up and begin swimming without having dedicated commands to do so. Imagine headphones that were in a knot, but could undulate so that the knot disappeared just by each part undulating. In our world, we think from the top down. We would take the ends of the headphones and thread them through the loops until the knot was untied. But if each individual part of the headphones had its own oscillator, knots and other configurations would be simple to undo. With a CPG, initial conditions no longer matter.

When Auke placed the robot into a small swimming pool, it swam continuously. The robot was resistant to perturbations that struck any part of its body. If it brushed up against a wall, its traveling wave was briefly disturbed before the robot bounced off the wall and swam away. This is similar to when you trip on a rock while running. You may stumble for a few steps, but your CPGs naturally return you to your original gait without conscious effort. Similarly, since the robot CPGs spontaneously produced their own signal, the brain did not have to command the wave pattern to reemerge. The autonomous nature of the CPG allowed the robot to simply return to its swimming state without thinking.

Today's robots do not use CPGs but instead have a central brain that sends commands down the line of the body. Often this brain consists of a computer processor, and it may not be in the head. Nevertheless, the brain leads and the rest of the body follows. If the brain is damaged, by decapitation for example, the robot simply stops walking. In comparison, the distribution of the pattern generators in the lamprey makes it robust. No single module is critical for motion. The use of CPGs is common across many vertebrates. There was a chicken in the 1940s that was meant to be killed for dinner, but a poorly placed axe blow left the chicken without a head, only a brainstem. For months, Headless Mike, as he became known, continued to walk and flap its wings. The CPG of the chicken was so robust that it could still command the chicken to walk even without its eyes.

* * *

Auke was satisfied with Amphibot's performance in deep water, but he found that Amphibot struggled in shallow water, and could not get to shore. Auke's next challenge was making Amphibot more amphibious. He realized that the robot's challenge was the same that vertebrates faced when invading land from water in the late Devonian era, over 350 million years ago.

The first large animals to come onto land were the tetrapods, the four-footed. They evolved into the amphibians we know today, the frogs, toads, and salamanders. To walk on land, the salamander evolved legs that are almost comical. They are small like a wiener dog's, but instead of extending underneath the body, they protrude out from the salamander's flanks like 90-degree pipe fittings. They appear like a last-minute addition.

Auke introduced these legs to his lamprey robot using *whegs*. A wheg or wheel-leg resembles a single spoke of a wagon wheel. It is driven by a spinning motor, but it also makes periodic ground contacts like a leg in order to step over small obstacles. To give life to this robot, he had to determine the coupling parameters of the CPG, this time from *Pleurodeles waltl*, the red-eyed newt. These salamanders were raised by Jean-Marie Cabelguen, a neuroscientist at Bordeaux University and fellow member of the Brotherhood of the Lamprey. The salamander's body shape is long and slender, similar to the lamprey. This meant that Auke could easily modify his lamprey robot. In fact, when the salamander swims through water it looks almost exactly like the lamprey. It tucks its legs into its sides and wiggles its body into a wave, pushing the water behind it.

Jean-Marie worked with salamanders about as long as a person's hand. First, he watched the salamander's motion both on land and in water. On land, it used a leisurely gait, its body wiggles at one cycle per second. As soon as it entered water, it doubled the pace of its wiggles. Moreover, it changed the shape of its wave from a traveling wave to a *standing wave*.

A standing wave appears when you use a jump rope. The endpoints of the jump rope are *nodes*, points that remain stationary. The points in between simply rise and fall, by various amounts, with the range of motion in the center of the rope referred to as the wave's amplitude. The salamander's body has nodes fixed at two places, the shoulder and hip girdles. The backbone in between oscillates laterally. In addition, the salamander steps its feet in time to the motion of its body.

To understand how coordination of its body and legs occurred, Jean-Marie anesthetized the salamanders and opened up their nerves in both the spine and the legs. He found neural oscillators in both places. The intrinsic frequencies of the leg CPGs were half as fast as the spine CPGs. Moreover, when the leg and spine were connected, the strong coupling between the two created a standing wave pattern just like the body motion when the salamander swam.

Jean-Marie then connected an electrode to the brainstem. He used a signal generator to send signals down the spine, mimicking the brain. At low levels, the stimulation leads to a walking gait. Because the leg CPGs had a much lower intrinsic neural frequency, he easily saturated them by increasing the frequency or intensity of the stimulation. At high frequencies of simulation, the salamander started performing fictive swimming. A similar behavior is found in a live salamander. On land, it simply walks. If you chase it, it increases the stepping frequency of its legs in order to break into a run. Once you grab its body, the salamander panics, increasing the frequency of its legs so much that the leg CPG saturates. Its body then generates standing waves as it is tries to wiggle out of your grip.

Auke programmed the intrinsic frequencies of the salamander CPG into his robot (Fig. 7.3). He could increase the drive to the CPGs and make the robot spontaneously go from walking to swimming. It was as

FIGURE 7.3. *Salamandra robotica,* a salamander robot that can both swim and walk. Control of its locomotion is reduced down to simply modulating the frequency of its brain signal, which induces the robot to undulate its body to swim to shore, and then crawl on land and walk away. This robot is controlled by central pattern generators (CPGs), networks of oscillators that spontaneously create waves of electrical activity along the robot. The waterproof robot is equipped with eight motors for spine undulations, and four motors, one per leg, for leg rotation. Courtesy of Auke Ijspeert.

if the robot's control of movement had been reduced to a stick shift. Low gear meant walking at low frequency, medium gear meant walking at higher frequency. Increase the gear even higher, and the robot would transition from walking to swimming.

This transition was not instantaneous. A brief resynchronization step of about half a cycle was required, during which the robot would do a half-walk–half-swim while the CPGs became aware of one another and resynchronized accordingly. As long as CPGs can send signals to each other, and they are given enough time, they can synchronize. This phenomenon has long been known but was made famous by Dutch scientist Christiaan Huygens, the inventor of the pendulum clock. If two such clocks are hanging on a wall, the small vibrations of each of the clocks travel through the wall and gradually prod the clocks to synchronize. This is one way that clocks in antique stores can all chime in unison.

Previously there had been no well-established methods to control robots with many independently moving parts like a backbone and multiple legs. Here, the CPG method worked well, and this was only the beginning. In animals, CPGs can be used to transition between very different gaits. Scientists have shown that a goose whose brain was detached can be stimulated in its brainstem to walk, run, and ultimately stretch out its wings and fly. Keep in mind that this transition between gaits involves independent muscles at different parts of the body, from the wings to the legs. Using CPGs, the transition between gaits can be accomplished quickly, which is useful in situations where an animal is being chased.

Auke believes that in the future, robots programmed with CPGs could become much simpler to control. The natural rhythm of walking would be engrained in the CPGs themselves. Since walking would be the natural default state for such robots, they would need a part of the brain to inhibit locomotion, as is done in vertebrates. Behaviors would not need to be separately programmed but could naturally arise from the CPGs. In the salamander robot, walking can occur no matter the initial configurations of the robot's body. Separate controllers are not needed for the legs. Instead, the legs are turned on and off simply by changing the brain's driving frequency, or the robot's state of alarm. Low frequency is a relaxed state, and high frequency is excited or frightened. These states cause the robot to transition from walking to swimming. The system would also be highly robust, which our current computer systems are not. For example, once its software gets a bug, a computer can freeze and stop working. Not so with the CPG-controlled system. It is highly robust to initial conditions and perturbations because its controller is everywhere and cannot easily be hijacked.

* * *

The nervous system uses CPGs that sense each other and generate signals. What if each CPG had legs and could explore the world? It would be like a swarming brain, each part of which could sense its surroundings and react accordingly. This is the strength behind swarming animals such as flocks of birds, schools of fish, and colonies of bacteria. In the next chapter, we consider how the many become one.

CHAPTER 8

Are Ants a Fluid or a Solid?

It was the Tuesday after Memorial Day weekend when I received a distress call from a neighboring professor. When he had arrived that morning, he sat down at his desk as usual. Underneath his desk, his foot touched something soft, like a sponge. When he looked down, he saw a miniature Eiffel Tower, about a foot tall, red and seething. It was a tower of ants. Not just any ants, but fire ants, known for their venomous sting. How did the fire ants manage to escape my lab?

On typical days, the fire ants were kept in my lab in plastic bins the equivalent of ten stories tall, their walls coated with liquid Teflon to make them into an unescapable ant Alcatraz. However, the three days of the long weekend had given the ants sufficient time to gather their colonies into an aggregate of 100,000 for a mass migration. They piled themselves on top of each other to build a siege tower to escape their bins. They carried their eggs and brood with them, laying chemical trails that would lead them across the lab bench and down the sides of the cabinets. They then proceeded across my laboratory floor, beneath my lab's closed doors, and then underneath the neighboring professor's door. Because of their ability to escape, we checked on the ant bins daily, but came to dread three-day weekends, which had over the years given me the title of most hated professor in the biology department.

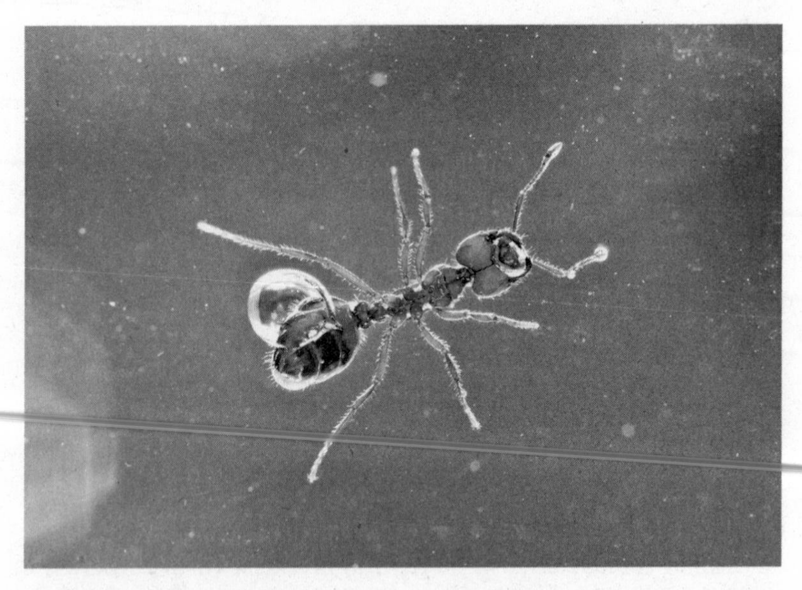

FIGURE 8.1. A fire ant submerged underwater, with its plastron, or artificial gill made of bubbles, clearly visible. Although ants are denser than water, they float by virtue of carrying a layer of bubbles on their surface. As ants breathe through their skin, oxygen in the bubble is depleted and then replenished by diffusion from the surrounding water. Their rate of oxygen use equals the rate of diffusion, allowing them to breathe indefinitely underwater.

Fire ants are not just capable of escaping from a lab. Years ago, they escaped an entire continent. Fire ants evolved in the Pantanal of Brazil, a broad flat wetland. During the dry season, the ants live in tunnels underground. But during the rainy season, 80 percent of the Pantanal is flooded, causing a number of animals to escape to the few remaining dry areas. The fire ants also evacuate, by linking their bodies together into a patchwork quilt. Their bodies can trap bubbles, enabling them to breathe indefinitely underwater (Fig. 8.1). They have sticky feet that they use to cling to one another. A colony of 100,000 ants forms a raft about the size of a dinner plate (Plate 10). During floods, a series of colonies can link together, like red seaweed on the water surface. Because they are so good at surviving natural disasters, they are spreading worldwide. In the 1930s, the fire ants had arrived in the United States through mysterious means, probably a shipment of potted plants that landed in the port of Mobile, Alabama. Fifty years later, the fire ants occupy a region over 3.6 million square miles in the United States, triple the size of the original thirteen colonies. They have also made their way to Europe, Africa, Asia, and Australia, easily colonizing regions close to the equator. They generally prey on invertebrates, but when pressed, they prey upon livestock, wreaking over $750 million of damage annually in the United States. Humans are generally safe, but residents of nursing homes have succumbed to fire ants that crawled into rooms and up bedposts.

Although they are newcomers, fire ants are most certainly here to stay. Killing them one by one would be futile. A colony consists of over 100,000 individuals, and it can only be stopped if one finds the queen, responsible for producing new offspring. This queen safely resides in underground tunnels several feet below ground, out of the reach of most insecticides. Applying insecticide to every ant mound across the thousands of square miles that they occupy would be far too costly to be feasible, leaving no known solution for eradicating the world's fire ants.

I first read about the fire ants while working on my PhD thesis on water-walking insects. Boston was too cold for the fire ants to invade, so my first sighting of fire ants waited until I moved down to Georgia for my faculty position. I met biologist Mike Goodisman, who had been studying fire ant genetics for years. Mike offered to take me ant hunting and I agreed. I recruited a new graduate student and avid mountain climber, Nathan Mlot. The three of us jumped into a biology department van and began driving.

On the Georgia Tech campus, pesticides keep the fire ants at bay. However, on isolated highways an hour north, unmarked by GPS, is a land claimed by the fire ants. Bordering the asphalt, in areas too hot for other ants to live, are the telltale mounds of dirt created by fire ants.

Beneath these mounds are an intricate network of tunnels, including brood, eggs, and the queen, all of which I had to obtain if I wanted a viable colony. I dug with a shovel, picking up over 50 pounds of dirt and ants and packing it into a series of buckets. Bringing the buckets full of dirt back to my lab, I was mystified by how to separate the dirt from the ants. With over 100,000 ants, I would never finish if I decided to separate them by hand. Luckily, Mike showed me an old trick employed by labs that study fire ants. He connected a hose to drip water drop by drop into the dirt-filled buckets. The slow dripping simulated rainfall, giving the ants ample time to rally their troops to the surface, just as they do in Brazil during the rainy season.

I returned to the lab the next morning, after the water had been dripping all night. Brown water filled the buckets and the soil was now submerged. Floating on the water surface was an amorphous object like an amoeba, but the size of a small dinner plate. Poking out from the edge were legs and antennae moving about. The raft was wide and flat, and consisted of two layers. On the top layer, ants walked atop other ants as if they were on land. On the bottom layer, ants held each other's legs, forming a dense weave. They lay still as if in a trance.

When I pushed down on the center of the raft, it deformed and ricocheted like a piece of rubber, while the surface of the raft remained dry. The spacing between the ant legs was so tight that water could not penetrate it unless the ant raft was pushed deep underwater. In Chapter 1, you saw that hairy water strider legs are non-wetting because of the large surface area that they present. Similarly, ants link their legs together to present a rough surface composed of entangled legs that trap air bubbles. For the ants, cooperation is absolutely necessary because the ants themselves are denser than water and would otherwise sink.

When I put on a pair of latex gloves and picked up the ant raft, it felt like a material I had never touched before, like a liquid yogurt that also had a lot of holes in it, like a pile of salad greens. Like a pile of salad greens, the raft was springy, and if I squeezed it down to a fraction of its height, it recoiled back to its original shape. If I pulled the raft apart, it stretched like cheese on a pizza (Plate 11). The raft could compress and stretch because it was connected by thousands of ant legs, which each could bend or extend its joints. But how can the raft be both solid-like and liquid-like? This is where the analogy with the salad greens has to be stretched a little: unlike salad greens, the ants have their own energy source and are highly active. I used a high-speed camera to film the ant raft apparently sitting still. Each of the ants' legs were quickly making and breaking new connections as if the ants were playing a high-speed game of tag. It happened so quickly that the structure appeared to stay the same shape, although it was continuously being composed of new bonds. This activity allowed me to roll the raft together in my hands like dough, forming a hot dog or a pretzel. As I did so, the ants were not injured, but they simply reconfigured their bodies. If I broke the raft into two, the ant raft could, without my help, reconnect when I simply put the two ends together.

The ability to reconnect like this is called *self-healing*, and it is a long-sought-after property of materials to make them more robust under forces and the passage of time. Self-healing materials are not yet on the market, but prototypes involve concrete imbedded with small pockets of glue that extrude if the concrete cracks. Many of these materials are inspired by human skin, which also self-heals through the clotting of blood and the building of new cells. What sets the ants apart from even self-healing materials is the speed that they can self-heal. Cuts in our skin take days to heal, as the body sends specialized cells to the region from long distances away. In contrast, ants can reform connections in fractions of a second.

If you place a penny on the surface of the raft, it is engulfed as ants begin climbing on top of it. Ants under the penny clamber to get out of the way, while more ants get on top of it, pushing it down even more. Slowly it sinks through the raft like a spoon falls through yogurt. The ants can do this because they do not have to be careful in how they connect their bodies; they can do so in nearly any orientation. We flash-froze and CT-scanned one of the ant balls, and found that an ant's legs attached to its neighbor nearly anywhere on its body. This ability to attach quickly and non-selectively allowed them to patch holes that the penny left behind, making the ant raft behave like a liquid. In a single drop of water, there are over a sextillion molecules of water, the number 10 with 21 zeros behind it. An ant raft has only thousands of links, yet it appears that this is enough for the raft to exhibit liquid-like behavior.

The ability to flow like a liquid or spring back like a solid is useful to the fire ants in a flood situation. Imagine an ant raft floating in the middle of a flash flood. It floats at the mercy of the current, and simply crashes into any obstacles that it encounters. If the currents are fast, the raft may be struck against rocks or other hard objects. The raft's springlike ability helps to cushion the ants and prevent injury. On the other hand, if currents are moving slowly, the raft can use its liquid-like ability to anchor itself. Ant rafts are often found lodged between weeds or other emerging vegetation. By flowing through such crevices, an ant raft can reach out to shore like an amoeba reaches out a pseudopod.

To further study the material behaviors of ants, I paired up with a local expert, Alberto Fernandez-Nieves, who has long studied *non-Newtonian fluids*, materials that are fluid and solid at the same time. To study such fluids, Alberto and his student Mike Tennenbaum use a rheometer, a high-precision blender that costs as much as a car. When it is on, the rheometer floats on air bearings, a thin layer of air that makes it nearly frictionless and able to perform extremely precise measurements, even if the forces are applied by ants.

Rheometers were invented by the chemical engineers who design fluids for our everyday lives. Shampoo, for instance, used to be available in powder form, which was difficult to apply to hair uniformly. A purely watery form was also inconvenient because it would flow through one's fingers. The engineers added long chains of polymers to transform the shampoo into a non-Newtonian fluid, a fluid that changes its behavior

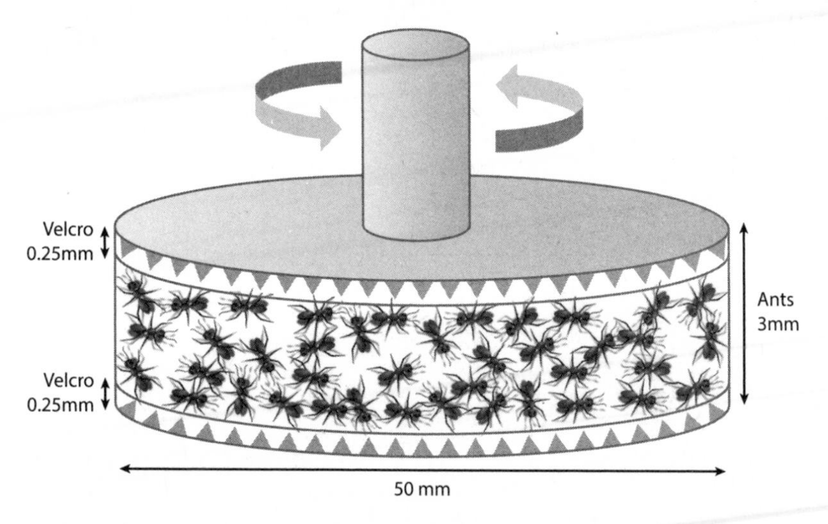

FIGURE 8.2. Rheometer apparatus for measuring the viscosity, or resistance to flow, of the ants. Torques are applied to rotate the top plate, and sensors measure the rate of rotation, due to the flow of ants within. Velcro is applied to both plates to reduce slippage between the ants and the plates.

depending on what you do to it. Long chains of polymers in the shampoo are arranged randomly when low forces are applied. Upon the application of high forces, the polymer chains align, reducing their resistance to flow. As a result, the shampoo sits like a chocolate chip in your hand, but easily smooshes into a puddle when applied to your hair. Paint is another example of such a non-Newtonian fluid. It flows easily when pushed with a paintbrush, but when applied in thin layers to walls, it solidifies, holding in place.

Ants tested inside a rheometer looked like they were in a small merry-go-round (Fig. 8.2). We fabricated a glass tube so the ants would stay inside without escaping. The tube had to be very carefully dimensioned to leave a small gap with respect to the top plate, which would be churning the ants like a blender. Any contact between the tube and the plate would introduce error into our measurements. A vacuum cleaner was kept nearby, which was our fire extinguisher if fire ants started escaping. When we measured the viscosity of a cluster of ants, we found that they had similar properties to paint. They held fast at low forces, but when the forces became too high they released their grip to prevent injury.

The rheometer test was useful because it gave us thresholds for applied force that would cause a mass of ants to transitions from a solid to a liquid. We could fit about 3000 ants in the rheometer, and by dividing the applied force in the rheometer by the number of ants, we could calculate the forces applied to a single ant. It is as if we were calculating how water would change from ice to liquid form by estimating the forces applied to a single water molecule. An ant mass would liquefy if, on average, a force of two ant weights was applied to a single ant. This threshold corresponded closely to the height of an ant raft, which was also about two ants. This number explained why ant rafts, initially at arbitrary heights, would always liquefy and settle to two ant heights. The independent measurement of the ant's liquid-solid threshold gave us confidence that we were measuring things correctly.

We began ant raft experiments by rolling the ants into a ball with our fingers, as if we were kneading dough. We did not want to overwork the ant dough. We wanted there to be air pockets between them. About 10,000 ants make up a ball the size of a small meatball. When this ball is placed on a water surface, it slowly spreads out into a pancake, like a dollop of ice cream melting. Our rheometer experiments have shown that the ants melt when the forces are too high. Initially, when the ball of ants is placed on the water, the ants on the bottom feel the weight of those on top, and begin releasing their grip. As the raft decreases in height, the ants continue to release their grip until the raft is only two ants tall. Then, the ants stop releasing each other and solidify. The liquid behavior enables the raft to change shape, yet the solid behavior enables the raft's final shape to stick. Without the last solidification step, the ants would disperse and lose each other on the water surface.

Clearly, the ant's threshold was critical to building rafts. But how would it affect other shapes, like towers?

Building a tower along a slippery surface brings to mind the annual initiation put to the entering plebes at the US Naval Academy. On campus is the Herndon monument, a granite obelisk over 21 feet tall. Every year since 1959, the obelisk has been greased with lard and the plebes asked to place a naval cap on the peak. The challenge is that a person can only tolerate one to two body weights atop her shoulders for long periods of times, yet the tower is at least five people tall. It is thus not possible for five plebes to stand on each other's shoulders to get the top because the bottom person would be supporting four plebes, and be beyond her limit. Instead, they must work together to redistribute the weight. This can be done by having each layer have an exponentially increasing number of plebes. For instance, the top layer can have one plebe, but the second layer must have at least two who equally support the weight of the top plebe. This means the third layer needs at least five or six plebes to support the combined weight of the first two layers. You can see that the layers increase in size very quickly. The plebes accomplish this tower shape by careful planning that is not possible for the ants. The ants build by distributed control, meaning that no ant is in charge of the task. Yet, the ant towers looked very similar to the one built by the navy plebes. This is where the ant liquefaction becomes important.

When we watch the ants build towers, we see that they do so by a process of trial and error. Sometimes the ants build a tower that is too tall and skinny, with a narrow base. That tower simply collapses, bending like a piece of Jell-O. This is because the ants at the base feel forces that are beyond their threshold and simply let go. The avalanches are the ant's way to return to the drawing board. The building of a structure like the Eiffel Tower was simply a matter of breaking down everything that put excessive forces on the ants. The shape emerged because it was the only one that could stand the test of time.

Fire ants are not the only ants that can aggregate together. Army ants, a group of over 200 ant species, aggregate in ways that are even more sophisticated than the fire ants. To study the army ants, let's leave Atlanta and head to Panama.

* * *

It was dusk and New Jersey Institute of Technology biologist Chris Reid and his friend George Washington University's Scott Powell had already walked the same small muddy footpath several times. They were on Barro Colorado Island, a large island in the Panama Canal, tracking the army ant *Eciton burchellii*. They swung their flashlights back and forth, scanning for the telltale foraging trails, up to 12 ants in width. The last time Scott was here, he had observed the longest bridge of ants he had ever seen (Plate 12). Completely composed of ants, it was the length of his arm, the equivalent of a human suspension bridge 90 feet long. The ants in the bridge held their bodies still, only their antennae flailing about wildly as they sensed the traffic moving atop the bridge. How could the ants build such a long bridge?

To increase their chances of spotting an ant bridge, Chris and Scott looked for the army ants' overnight encampment, the bivouac. The bivouac was 700,000 ants strong, and contained the same number of hungry brood within. The brood was hungry and growing, collectively requiring 40 grams of dry protein each day, roughly the size of a hamburger. To feed the brood, the colony had to get organized. It sent out foragers, whose mission would be to scout the forest, locate a food source, and then return the food to the bivouac as quickly as possible. The speed of the foragers was the rate-limiting step for the colony's growth. Small gains in their speed would compound over the lifetime of the colony, enabling broods to be raised faster, and more foragers to be sent out.

Years ago, Scott, along with his advisor Nigel Franks, had observed a remarkable method that ants use to increase their foraging speed. To an ant, leaf litter is not very flat, but instead resembles a road covered in potholes. These potholes would ordinarily slow down the ants, which would have to go around them. Scott observed that when an ant crosses an open pothole, it gauges the hole size compared to its own body. If the ant fits, it slots itself right in, grabbing onto the rim with its legs and becoming a manhole cover. Big potholes are filled in by large ants, small potholes by small ants. The range in size among colony members allows for all the potholes along a trail eventually to be filled in. The sacrificed ant is not injured, because the small size of ants makes them tremendously strong. An ant is only crushed by a thousand body weights, the equivalent of a house.

By momentarily sacrificing a single colony member, the colony gains tremendously. Every ant that uses the sacrificial ant as a stepping stone gains a second in traveling time. Consider a foraging trail 200 meters long, the equivalent of 22 miles for a human. The trail is 12 ants wide and the high speed of the ants results in 200 ants per minute crossing over the pothole. If every pothole of the road is filled in this manner, ants double or triple the rate they can find and bring back food. This is just one example in which small changes in behavior, when multiplied by the numerous ants in the colony, can have huge savings for the group. Later in this chapter, in our story about robots, we will see this same type of multiplied behavior arise not just with benefits but also with consequences. When one robot makes an error, over multiple robots these errors can become real threats to the swarm. A group's large size magnifies benefits and consequences equally.

The pothole-filling behavior, however, could not explain how bridges were formed, especially bridges over 15 ants long. Scott had a hypothesis. Perhaps the bridge was only built along certain structures, such as a Y-shaped tree branch. To cross the prongs of the branch, ants traveled down one prong of the Y, then back up the other prong. The fork of the Y was the bottleneck. If a single ant lodged itself there, it could reduce the travel distance incrementally for the colony. Over time, that ant would be joined by several others, building a short bridge. By a process of accretion of ants on one side of the bridge and removal of ants from the other, the bridge lengthened, and migrated

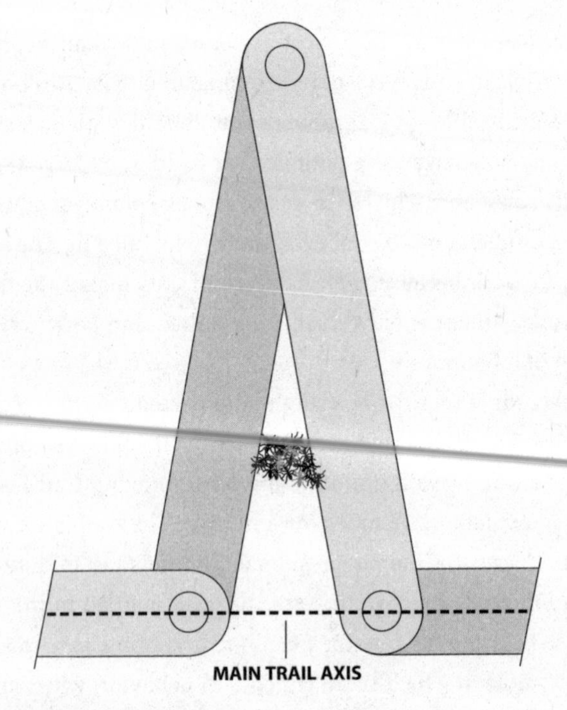

FIGURE 8.3. W-shaped apparatus for observing construction of army ant bridges. The segments are made by 3D printing, and the circles represent hinges. The entire apparatus intersects the trail of ants, and the ants are forced to walk along the apparatus to return to the trail. To reduce their travel time, ants build bridges at the top of the apparatus, and over time, the bridge migrates downward.

down the prongs of the Y. In other words, ant bridges could not appear out of nowhere. They had to be seeded, and then they needed time to migrate.

To test this idea, Chris 3D-printed four platforms, each similar in size and shape to a ruler, and arranged them like a W (Fig. 8.3). The platforms were hinged at the endpoints, allowing the W to increase in width, challenging the ants. They filmed the ants building bridges across the W. As Scott had first imagined, the ants started by going the long way around the platform, tracing the shape of the letter W. As traffic increased, the ants' path began to migrate toward the crevices of the W. Finally, a few ants ventured out and crossed the crevice, holding their bodies out as pothole covers. Over a period of hours, they formed a bridge that migrated toward the center of the W. Scott and Chris cheered as they watched Scott's hypothesis come true.

After an hour of watching, Scott and Chris stopped celebrating. The bridge had stopped moving before hitting the center of the W. This behavior occurred for every bridge they observed. It was surprising, because the ants would get much more benefit if they could walk straight across the W. Instead, the bridge had stopped migrating before it was of maximum benefit. If bridges were meant to reduce travel time, why would the ants go to all this effort only to stop short?

The answer to the question lies in thinking from the perspective of the entire colony. From the colony's perspective, each ant is an individual that should be used to best increase the flow rate of food back to the colony. Thinking in this fashion, relegating a few ants to making up the bridge is not a sacrifice for the colony. It helps the other members get to their destination faster. However, this task allocation also has consequences. The ants in the bridge can no longer bring food back. A bridge that migrates lengthens and requires more ants to sacrifice themselves. The final position of the bridge represents a trade-off. A bridge closer to the main trail axis reduces travel, increasing the rate of ants returning with food. However, the movement of the bridge also entrains more ants into the bridge, reducing the number that are foraging. The reason the bridge never intersected the main trail was that it began giving the colony diminishing returns.

If the ants could only communicate to their neighbors, how could they optimize the position of a large structure such as a bridge? The position of the bridge influences its length and, in turn, the density of free ants on the foraging trail. If the bridge is short, there are more ants along the foraging trail, bringing food from its source to the bivouac and back again. This creates a greater density of ants around the bridge, and increases the chance that one of them will join the bridge. Now, let's consider if the bridge is too long. In that case, the number of ants on the foraging trail is low because too many are entrapped in the bridge. Each ant in the bridge counts the footsteps above it. When entrapped ants feel traffic atop them is lower than a desired set point, they leave the bridge. The bridge was thus dynamic and sensing. It migrated according to how many ants were on the road, and automatically moved to the position that was best for the colony. Imagine that in a traffic jam, cars voluntarily went to the side of the road to widen the roadway for others until the traffic jam was over. The internet of things and an increasingly connected world may lead to an age where objects begin cooperating for our benefit.

In our last examples, we saw how a swarm's size can give it a sixth sense for its surroundings. This ability of swarms did not come about overnight; it was honed by evolution over millions of years. In our next story, we will see how to build a swarm from scratch.

* * *

Harvard University computer science postdoc Mike Rubenstein and his advisor Radhika Nagpal had finally made their quota of robots for the month. Thirty robots had been built by hand, and surprisingly, they all worked. It was time to upgrade that number to one thousand, and give the robot model a new name, Kilobot. Kilobot was the size of a cookie, and looked like a miniature barstool with a circuit board sitting on top. It cost $14 to build, which seemed small, but ordering the thousand replicas would set Mike back $14,000, the cost of a small car. With shaking fingers, Mike typed in the request to order the materials online. He had done his PhD thesis on computer simulations of cooperative robots. For the last 10 years, he had been pondering how robots in a swarm would think, talk to each other, and get out of trouble. Now he would put these ideas to the test.

Robotics is a scientific field only 50 years old. Since the 1980s, robot designers have dreamed of a modular robot consisting of smaller robots sitting in a container such as a bucket. Each of these smaller robots would have its own small brain and an ability to propel itself. When you needed to call on the robot, you would simply just empty the bucket out. The modules would link together to build a larger, more capable robot. A robot made of modules would have several advantages. Modules were expendable, and could be easily replaced if damaged. This modularity would also be useful in places where humans cannot go, such as in space. Imagine taking a number of trips to a space station, each time bringing along robot modules that would be able to combine their efforts together with the others already there.

For modular robots to become a reality, each module must be made simply and cheaply. When Mike was writing his doctoral thesis on swarm algorithms, he saw robot swarms created on the scale of 10 to 50 robots. He realized that true swarms could not be formed with so few robots. In his thesis, he wrote algorithms where swarms could self-assemble and self-heal. In self-assembly tasks, his robots joined together to form a desired shape, like a starfish. The quality of self-assembly depended directly on the number of robots. With only 10 robots, he could only build simple shapes, like a circle or a square. With 100 robots, he could begin to build more sophisticated shapes, such as a wrench or a key, tools that such robots would be able to form, and we could then use. Mike found that to make those tools look at all like their original counterparts, he needed at least 1000 robots. Even that number of robots is far less than the pixels on a computer screen, but it would be a start.

As Mike began designing his swarm of 1000 robots (Fig. 8.4), he started to see interesting problems occurring due to the large numbers of robots involved. Imagine you had just bought a new car with an excellent factory rating. Every hour of driving, it had only a one-in-a-million chance of a critical engine failure. Such a low chance would make the failure undetectable during testing of a single car, because it would probably take up to 100 years for it to show up. However, if you had a fleet of 1000 cars running simultaneously, an engine failure would likely occur in the first month. Rare problems that emerge over impossibly long times for an individual become inevitabilities for a swarm. Therefore, it seemed that he would have to make the robots perfect if he wanted to avoid problems.

To make the swarm affordable, he also had to tackle the inevitable trade-off of quality for quantity. If he were building a single robot, he could spend hundreds or even thousands of dollars on finely tuned sensors and highly precise locomotion abilities. The greater the number of robots he wanted to build, the cheaper and less precise they would have to be. The problem was that no one had written algorithms for controlling a horde of error-prone robots. As his robots became less dependable, he needed new algorithms that could make up for the robots' errors and still accomplish tasks robustly. Mike kept these issues at the back of his mind. He tried to imagine all the things that could possibly go wrong with the robots. He decided to start small, building 30 robots and testing them for very long times before he built the thousand. Once he pulled the trigger to build the thousand, he could only watch as his dreams and nightmares unfolded.

The robot would have to move about. Even a simple toy car's wheels were too complex if he had to build a thousand robots. A single car had too many parts: a motor, two axles, and four wheels. That was a minimum of seven parts already. Instead, he needed a device that could move with just a single part and no wheels at all.

The inventors of the cell phone could not imagine that the world would one day be filled with over a billion of the devices. The popularity of cell phones led to cheaply manufactured components, one of them the cell phone's vibration motor, which spins a small weight. This vibration motor led to a simple robot called the bristlebot, described

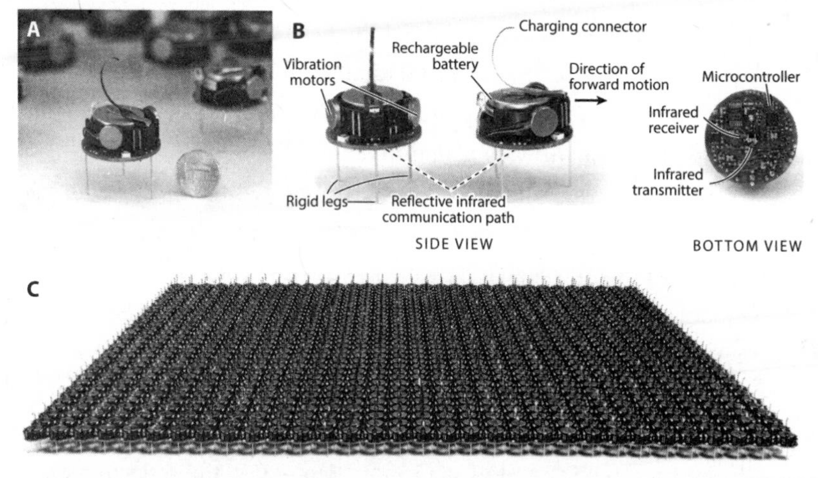

FIGURE 8.4. (A) Kilobot robot, shown alongside a US penny for scale. (B) Each Kilobot has a small onboard computer for executing programs autonomously, two vibration motors for moving straight or turning on a flat surface, and a downward-facing infrared transmitter and receiver. Robots communicate with their neighbors by reflecting infrared light off the ground. (C) A swarm of 1024 Kilobots. The Kilobot design allows for all operations on the entire swarm, such as charging or programming, to take a constant time to complete, independent of the number of robots in the swarm. Courtesy of Mike Rubenstein.

on evilmadscientist.com. Bristlebot is a vibration motor attached to a toothbrush. The vibration causes the bristlebot to glide forward, but it cannot change directions.

To build a robot that could turn, Mike attached two vibrating motors in earmuff position on his robot (Fig. 8.4B). The position of the two motors created a well-defined head of the robot. When the motors both rotated counterclockwise, the robot moved forward. When they rotated in opposite directions, the robot spun in place. The robot glided smoothly but slowly, with a speed of one centimeter per second, or 0.03 mph, the speed of a snail. As long as the underlying surface was smooth, the robot was now fully mobile. Each motor was only $1.50, making the propulsion system $3.00 total. He had to keep close tabs on the costs because he would be ordering a thousand of every part. This choice of motor had already committed him to $3000.

When a robot was surrounded by others on the table, it had to be able to speak to them. If the robot were circular, the most neighbors it could have in a hexagonal lattice was six. How could it talk to these neighbors all together? The cheapest way to do that was with light. Initially, Mike considered putting a beacon on top of the robot, like a police car, and small sensors, or eyes, on each of the surrounding robots. But that required the use of two reflectors, one at the beacon and the other on the receiving end of the robots. Moreover, the reflectors created losses that reduced the distance the beam could travel. Ultimately, Mike turned his idea upside down and used the table as the reflector. He attached a single infrared LED to the underside of the robot (Fig. 8.4b). If he suspended the robot on three metal stilts about an inch tall, the reflected beams could travel farther. A single eye in the center of the robot's belly allowed the robot to see the infrared beams from the others. Now the robot could speak and hear.

Mike needed a hands-free approach to working with the robots. Even an innocent on-off switch could spell disaster. It takes at most a few seconds to locate and touch a light switch. But five seconds multiplied by a thousand was an hour and a half. Instead, he needed to communicate to all the robots at once. He decided to connect a single robot to an infrared floodlight that was attached to the ceiling of the lab. Its beam would then send a code in the form of flashes to the entire army of robots.

He was able to buy all the parts in large quantities except for the charging wire, which he built by hand. To coordinate one thousand robots charging at the same time, he could not use individual cords. Instead, a small, stiff curlicue of a metal wire was attached to the head of every robot like a metallic haircut (Fig. 8.4B). A large metal sheet was placed on their heads, to charge them simultaneously. The only problem was with the sheet overheating if left on too long. Each robot took 1/10 of an amp of current, and the sheet a total of 100 amps, enough to run several appliances at once.

He began testing communication using a pair of robots. The simplest task that two robots can accomplish is an orbit, where a single robot is stationary, and another robot walks around, like in duck-duck-goose. It sounds easy, but that is because we use vision to guide us. Imagine playing duck-duck-goose in the dark. In that case, the person in the center shouts continuously. The orbiter listens to the shout. Based on the distance away, the intensity of the sound will increase or decrease. If the sound feels too loud, the orbiter increases the radius of its walking path. If it feels too soft, the orbiter decreases its radius. As we learned in the last chapter, this device is a called a *proportional controller* because the robot adjusts its behavior in a way that is proportional to the signal received. In this way, the orbiting robot traveled in a jagged yet roughly circular path.

Next, Mike worked with four robots, a sufficient number for the robots to determine their location with respect to one another. This concept, called *localization,* is imperative for working together as a well-coordinated group. To enable localization, he used ideas similar to a global positioning system (GPS). Your phone's GPS works by listening to four satellites orbiting around the earth. These satellites shout out regularly. Your phone receives signals from each of the satellites, and uses the timing of the signal to estimate the straight-line distance between it and each of the four satellites. The four distances are sufficient to calculate 3D coordinates in space, which is how your phone tells you where you are. Since Mike's robots sat on a table on a single plane, he only needed three surrounding robots to accomplish the same task. Mike had each of three robots shout out with an infrared signal. A fourth robot, called the sensing robot, listened for these signals and used them to calculate its coordinates. This robot demonstrated to Mike that it understood where it was by moving to the center of the three robots.

The localization worked well when there were only four robots. But with greater numbers, problems began occurring. One problem that arose was the cocktail party effect, where multiple conversations going on made it difficult for the robot to sense the signal from its neighbors. Mike used a common cell phone protocol called CDMA that makes sure everybody takes turn on the shared channel, like taking turns moving at a four-way stop sign. Moreover, the robots followed another rule of etiquette: only talk as loudly as necessary for the neighboring robot to hear. He decided that when a robot talked, only robots within a radius of 10 centimeters should be able to hear. Later, with additional numbers of robots, he would find that this protocol needed editing. When there were 1000 robots, there was a small chance of multiple conversations adding up, like the din of multiple clocks ticking in the same room. This ambient noise made a robot sense that its partner was talking louder than it was in reality, and thus under-predict the distance to the robot. To solve this problem, the robot would conduct multiple measurements to be sure of what it heard.

As Mike continued to build robots, he noticed that they were in fact limited machines with sensors that outputted noise, irregular fluctuations that obscured the intended signal. He began to hone his manufacturing process to reduce the noise. For example, he found that their infrared sensors were of varying sensitivity. After making many robots, he learned that the solder he was using was damaging a few of them. He switched to a low-temperature solder.

Finally satisfied with the design of the robot, he ordered the 1000-robot circuit boards to be made and delivered, and he assembled them over several months. When he was done, he arranged his 1000 robots in a rectangular phalanx. Over the next few days, he would test their ability to swarm together into preset shapes (Fig. 8.5), tasks that would take hours to complete. Their first task was to assemble into a simple five-pointed star (Fig. 8.5C). A cluster of four robots in the phalanx were designated as the seed of the star. This seed would remain

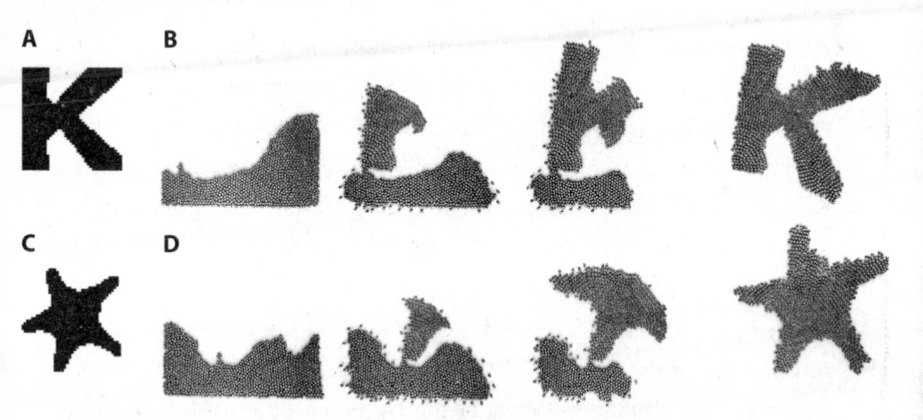

FIGURE 8.5. Self-assembly experiments using up to 1,024 Kilobot robots. (A and C) Desired shape provided to robots as part of their program. (B and D) Self-assembly from initial starting positions of robots (at left) of the letter K and a star. The final self-assembled shapes are on the right. Courtesy of Mike Rubenstein.

stationary, forming the bottom-most tip of the star. The other robots would rearrange themselves around this seed. A series of simple algorithms made it possible for the phalanx to change from a rectangle to a shape that closely resembles a star (Fig. 8.5D).

In the next step, the robots formed a gradient, a ranking of each robot in terms of distance from the seed. Robots next to the seed would have the rank 1, robots a little farther away would have 2, and so on. This task sounds easy when one has the godlike perspective of looking at them from above. But imagine if you were in a dark, crowded room and had to accomplish this task only by communicating with your neighbors. The gradient ranking was accomplished from the seed outward. When the seed shouted out, robots within 10 centimeters heard it and assigned themselves the number 1. Robot team 1 would then shout, and the teams within hearing range and that had not yet been designated became team 2. The process repeated until all robots knew their distance from the seed.

The gradient ranking set the order at which the robots could begin moving. The robots farthest from the seed would move first, then movement would proceed in order of ranking until the robots closest to the seed moved. Moving in order of the gradient ensured no holes formed between the remaining members of the phalanx. Holes were devastating to the swarm because a robot only knows where it is if it has neighbors. Once a robot is isolated, it loses its localization ability and falls off the map.

One by one, the farthest robots followed the edge of the swarm until they reached the seed. The seed knew where all the robots should go in reference to the seed. Since the seed in this case was the leg tip of a starfish, the remaining robots began filling in the rest of the leg. It took 12 hours for the robots to fill all five legs of the starfish. Only a few robots were moving at a given time, allowing most of the robots to rest during the test—important since the robots only had a three-hour battery life. A video camera recorded the robots while Mike slept. The next day he replayed the video to see how the robots did.

A few problems arose requiring Mike to repeat the experiment before it was successful. First, the motors, which had only cost one dollar, were highly variable. It was inevitable that as a line of robots followed an edge, a slow robot would be holding back a crowd. This occurs because the robots ahead of the slow robot would speed on ahead, leaving the slow robot as the bottleneck. He programmed the robots to sense who was ahead of them and then slow down to avoid collisions, just like in a real traffic jam. These traffic jams resulted from real-world variation in robot quality. Years before, Mike had spent his PhD thesis writing computer programs to simulate robots like these, but he had never suffered these problems because his simulated robots were all capable of traveling at the same speed. Working with large numbers of imperfect robots required special algorithms, teaching him things that a computer code could not.

Another issue that Mike had not anticipated was erosion, whereby the edge-following robots would push stationary robots out of place. By positive feedback, the more the robots were pushed out of place, the more collisions they would cause, pushing even more robots get out of place. After some time, these displaced robots created sharp edges, like rivulets dripping from fresh paint. Ultimately, shapes like the star and the wrench appeared warped, as if they were melting.

Since the time that Mike developed his 1000 Kilobot swarm, they have been used as an educational tool to study collective behavior of systems that are difficult to observe, such as the development of embryos. An embryo changes shape from a ball of cells to a hot-dog shape by migration of cells from one position to another. The robots can also be used to simulate the behavior of a range of processes in biology, such as *phototaxis*, movement toward light.

Modular robots throughout the world are capable of movement in a range of environments. Mark Yim at University of Pennsylvania works on modular robot algorithms to control cargo boats. On the ocean, the US Navy was interested in cargo boats that could link themselves together to act as landing pads for helicopters and bridges for jeeps. The main problem was the ocean waves, which tend to rock floating objects back and forth, and even bang them together if they are too close. Mark's solution was to build cargo boats 1/12 actual size and test them in a university swimming pool. The boats were rectangular and had twin propellers at the bottom to enable them to either spin or move in a straight line. Ropes on the edges of the boats mated with hooks on other boats. The same principles for Kilobot, such as seeding a structure to be built and building a gradient to fill in robots, were used for these boats as well.

Another interest in the field of modular robotics is to build structures in three dimensions. Termites can build mounds that are eight meters tall, equivalent to a human building a mile tall. The termites do this by carrying individual grains of dirt and laying them down one by one. Justin Werfel and Radhika Nagpal at Harvard University worked to build robots that pick up and place blocks. The robots could then climb up the stacks and lay more blocks. After placing dozens of blocks, a small castle was built. The challenge in this case was building the structure without the robots' bumping into one another. The structure had to be built in such a way that the robots could always climb it to build the next layer. Building in three dimensions has its own idiosyncrasies that researchers are overcoming, bringing us closer to the proverbial bucket of parts that can be poured down a drain, and ultimately unblock it from the other side.

CONCLUSION
The Future

I was at a conference when I received an urgent message from one of my university's reporters. In a grim voice, he told me to watch *Fox and Friends*, a daily morning show. When I turned on the show, a newscaster was speaking about the government spending millions of taxpayer dollars on research. She said, you might expect these funds to be spent on problems like AIDS, a cure for cancer, or the Zika virus. If these areas of research received more funding, then scientists might have found a cure for them by now. No, instead funds were spent on other kinds of research. She turned to a large game-show wheel labeled the Wheel of Waste, with the names of scientific studies on each of the panels. She was joined by a senator, and together, the two of them spun the wheel. As the wheel slowed, the flapper stopped on a familiar study: the wet-dog shake. They began an inquisition, a harsh criticism of why this work was a waste of taxpayer money.

They went over a few more of the studies: How many shakes does it take for a wet dog to dry off? Which has more hairs, a squirrel or a honeybee? How long does it take to pee like a racehorse? By the end of the show, I had become personally responsible for three of the twenty most wasteful studies for 2016. That's 15 percent of the nation's most wasteful research. My university later told me that as far as they knew, no one had ever been responsible for so many items in one year. While they were mortified, I was in some ways quite proud. I had never been responsible for 15 percent of an entire nation before.

After I had watched *Fox and Friends*, I turned to the senator's annual pamphlet of accused scientists, his "wastebook." This year's theme was called "Twenty Questions That Will Leave You Scratching Your Head." The book was written in fifth-grade language and had a picture of a *Tyrannosaurus rex* on the cover. I thought of my own children, Harry and Heidi, who were then not yet in elementary school. Soon, they would be able to read the senator's attacks, and they might wonder themselves, is research on animal movement a waste of money?

Scientists who study animals received a strikingly large proportion of attention from politicians. It's unfortunate for people like me, but it makes perfect sense. Before the wastebook can convince an audience, it must capture their attention. In 2015, Duke University biologist Sheila Patek was featured in the wastebook for building a "shrimp fight club" to observe the devastating underwater punch of the mantis shrimp. In 2011, biologist Lou Bernett built an "underwater treadmill for shrimp." The animal subjects of these studies certainly grab your attention. But if you spend the time to look deeper, it is clear that they are so much more than an obscure obsession.

As Shelia Patek knows, the mantis shrimp is an amazing animal. It accelerates its appendages faster than a bullet out of a gun, yet somehow its appendages remain undamaged. When propellers move that quickly, they generate cavitation bubbles, vapor pockets that explode and eventually ruin the propeller. Understanding how animals like the mantis shrimp move so quickly can lead to cavitation-resistant propellers.

Lou Burnett's treadmill for shrimp was actually an effective way to measure the metabolism of shrimp, an economically important species that is rising in consumption. Agricultural scientists have conducted similar studies of pigs, cows, and other farm animals. Understanding the rate of energy use of shrimp will play an important role in their farming and welfare. Judging a scientific work by its name is like judging a book by its cover, or a person by a single glance.

Basic research can be used in unforeseen ways. As you learned in Chapter 3, I performed a study of urination, showing that the shape of the bladder and urethra was responsible for animals urinating for constant times, about 21 seconds. Since my study was published, scientists from around the world, from Japan to the Netherlands, have cited the work as influencing their own.

When you go see a doctor with urinary problems, she must measure the flow rate of urine using expensive lasers and equipment. A doctor in Japan, Seiji Matsumoto, saw our study and thought that urination time could be a simpler way to detect urinary problems. He interviewed two thousand Japanese people, from children to people over 80 years old, about their urination time. He found that urination time increases with age, from 21 seconds at age 20 to 31 seconds at age 80. This increase in time is due to a combination of an expanding prostate and a decreasing strength of bladder muscle. One day, doctors may ask "How long did you urinate today?" as an initial way to infer bladder health before performing ultrasound testing.

Incontinence is a widespread problem for the elderly. One solution would be to implant an electronic prosthesis in the bladder that zaps it with electricity, enabling better control of the onset of urination. But a healthy baseline is needed to determine if proper bladder function has been restored. The baseline chosen by University of Michigan engineers Abeer Khurram and colleagues is my 21-second law of urination, which they use to test their prosthesis in cats to ensure it is safe for human implantation.

Organ regeneration is an important new field, and regenerating the urethra is no exception. Dutch scientists Luuk Versteegden and colleagues have designed a replacement for the urethra out of collagen and human cells. Like anything that is placed into the body long-term, it must first be tested for durability. The test chosen by engineers is three days of real human conditions, involving 21 seconds of urination every two hours.

These studies—all published in 2017 alone—show that the 21-second rule can provide critical input into designing treatments, prostheses, and artificial organs, work that directly helps people.

I started this chapter talking about science that is wasteful. The concept of waste is based on the notion of a limited gas tank and a single known destination. People expect scientists to save gas as they go from A to B. But the real power of science is to take us to destinations that we have never been to, like the discovery that urination time can be a marker for bladder health. Many of the other subjects in this book are similarly bizarre and unexpected, so that we might never have come upon them without scientists who study animal movement.

This book may serve as an extended answer to the attacks by the senator, and to questions as to whether animal motion research is worthwhile. Since I began writing the book, the field of animal motion has expanded. Therefore, I would like to spend this chapter discussing the future of the field. Research in animal movement has come a long way since Sir James Gray performed his first studies of fish swimming in 1939. He could not have imagined the robotic fish and other applications that are emerging less than a century later. What else can we expect in the next ten, twenty, or one hundred years? How will animal motion research change, and how can we expect it to change the world?

The way scientists take pictures is changing. In animal motion, photographs and film have always been an important way to obtain data. Until now, the photos have merely increased in speed or resolution. University of Massachusetts biologist Duncan Irschick is changing that. He has built a device called Beast Cam that can take three-dimensional pictures of living animals. The device consists of twenty inexpensive consumer-grade digital cameras hanging together like lights in a miniature stadium. The cameras take pictures simultaneously from different angles. A computer algorithm assembles these pictures together into a three-dimensional point cloud. I visited Duncan's lab last year and admired the yellow and black spots of a rare Panamanian golden frog he had taken with this technique. Unfortunately, the frog is nearly extinct, but at least Duncan had saved its 3D image digitally. Using his computer mouse, I could see different perspectives of the frog as if I were walking around the animal, and appreciate its fine textures and colors, as if the frog were thriving today. Even the posture of its feet and the arrangement of its toes were unique, making the image strikingly lifelike. Although it's too early to tell how scientists will use these 3D photos, Duncan's work has already attracted the attention of the entertainment industry. It is far cheaper to take a 3D picture of a frog than to have an artist render one by hand. One day movies and news articles alike may employ 3D animals.

There is also a potential downside to this increased realism. It may supplant people's desire to see real animals. Instead of going to the zoo, some may find it sufficient to see 3D recordings of animals in virtual reality. This is a real possibility because people of all ages are spending an increasing amount of time on their handheld devices consuming content from social media to news to movies. I, however, believe that virtual reality may help the conservation effort as much as color photography has done. For instance, tigers in the wild are decreasing in numbers, with more tigers in captivity than there are in the wild. The public's awareness of tigers is made possible through nature photography and documentaries. Imagine, how many more people would become aware and supportive of conserving animals if they could see them in 3D?

Beast Cam is one view of the future, where components of cheap consumer technology are bootstrapped together to produce something new. Much of the work in our field is done this way, with homemade equipment that the Department of Defense calls "arts and crafts." With the decreasing cost of 3D printers, homemade equipment is going to become even more popular. At the same time, high-tech equipment is also pushing the envelope for capturing images.

Steve Gatesy and Beth Brainerd at Brown University are one of the first teams to use 3D XROMM, or X-ray Reconstruction of Moving

Morphology. They first CT-scan an animal to determine the shape and position of its bones. X-ray cameras then track the motion of the animal's bones in 3D as it runs, feeds, or performs other motions. The combination of the two tools gives scientists 3D X-ray vision. The result is a 3D video of the animal's skeleton racing across the screen. By transplanting markers into muscles, scientists can track the contraction of muscles along with the motion of the bones, which helps us to understand how muscles produce motion. Such techniques are important for understanding complex movements involving many parts, a regular occurrence in nature. When we speak, we use a number of separate parts in our face—43 muscles and 14 bones, in fact. Researchers can use 3D XROMM to help understand the roles of these different parts, and how they can be repaired to restore function in injured patients.

Three-dimensional imaging can also help make museum collections of animals more accessible to the public. For the last few decades, research programs like the Digital Morphology have established websites like digimorph.org that make available terabytes of skeletons of both dinosaurs and existing animals. As CT-scanning technology becomes more affordable, such efforts are expanding. Adam Summers at the University of Washington is head of an initiative to "scan all vertebrates," to generate 3D images of over 20,000 birds, fish, reptiles, and mammals. These specimens are currently on shelves in museum collections. Adam's work will make them available online and free to all. Moreover, the scans will reveal not just bones, but also the entire cardiovascular system and nervous system, vast networks that span the entire body. Roboticists are already using the data from Adam's scans to design limbs for underwater robots.

Animal motion research is going to involve a flood of digital data. If a picture's resolution is doubled, the data goes up by a factor of four. But if a 3D picture's resolution is doubled, the data goes up by a factor of eight. This increase in data is inevitable as sensors become more accurate. Indeed, the challenge in the future is not going to be getting data as it was in the past, but developing new ways to interpret, store, and communicate it to others.

In the future, the science of animal motion will also become more integrated with fields like developmental biology, the study of the growth of organisms. Developmental biologists are mainly concerned with model organisms such as the small nematode called *C. elegans*. This organism reproduces rapidly and can be easily raised in the lab. Moreover, 35 percent of the nematode's genome is closely related to the human genome. This similarity makes the nematode an ideal test subject for the effects of treatments and drugs that are planned for humans. To determine the effect of drugs on *C. elegans*, scientists often observe them in different environments. University of Pennsylvania mechanical engineer Paulo Arratia has been observing *C. elegans* in fluids of different viscosities as assays for swimming strength. Swimming strength can be in turn used to quantify the effects of aging or drug treatments on the *C. elegans*, discoveries that will ultimately be used on humans. Understanding animal motion is critical to interpreting such tests.

Animal motion will be increasingly important to the study of agriculture as the world's population grows. Since humans are likely to continue to eat animals, studying animal movement has an important role to play in animal welfare as well as the agricultural bottom line. London-based biologist John Hutchinson works on the biomechanics of a domesticated chicken that he calls "chicken of the future." The problem with the domesticated chicken is that it is bred to grow its breast so quickly that its leg muscles cannot keep up and eventually the chicken is unable to walk properly. For the last segment of its life, it is too large to stand up and it sits, unable to move. John X-rays chickens while they walk to analyze their gait. He sees a future where biomechanics works hand-in-hand with farmers to breed chickens in a humane way.

With my graduate student Olga Shishkov, I work with Grubbly Farms, a startup founded by Georgia Tech students. This farm has no cattle or corn. Instead, it is full of black soldier fly larvae, small grub-like insects that are a prime candidate for the sustainable protein source of the future. These larvae can consume food quickly, and are also not picky about what they eat. Currently, restaurants and homes in the United States generate an unimaginable 1–2 billion metric tons of food waste per year. The idea is that these grubs would feed on the waste and then the grubs themselves would go toward feeding chickens, fish, and other agriculturally important animals. It's a win-win situation because currently food waste ends up in landfills and eventually contaminates groundwater. The problem with raising these larvae is that that optimal feeding and housing strategies are not yet known. Unlike chickens and sheep, larvae can be packed in three dimensions like peanuts. We are working to see how their motion affects their feeding rates, and how the shape of their housing affects the maximum numbers that can be kept. It's part of a growing area of research called *active matter*, where physicists are trying to predict the motion of swarms using rules similar to thermodynamics.

In this book, we have already seen the influence of robotics on animal motion, including robots inspired by water striders, jellyfish, sandfish, and others. Another area of increasing growth is soft robots. Such robots are inspired by the bodies of caterpillars, trunks of elephants, or arms of octopuses. One of the applications of such robots is to move safely around people, especially the elderly. In Japan alone, people over the age of 65 are currently 25 percent of the population, but the proportion is predicted to reach 40 percent by the year 2060. Current industrial robots, such as the ones used in car factories, cannot meet these needs. Such robots have rigid segments separated by high-torque motors, and are designed to work repetitively in predictable environments. The environments around humans are far too unpredictable for such robots. Soft robots are the only candidates for a robot that can brush a person's teeth or comb her hair. If such a robot makes a mistake, no one is hurt.

Soft robots are being built from an increasing range of materials. A soft robot has been made in the shape of a manta ray using genetically engineered rat muscle tissue that contracts with flashes of light. Such soft robots have muscles that are grown within silicon templates, and are thus composed of both animal cells and built parts—*biohybrids*. Other soft robots combine the anatomy of animals with origami. A robotic snake uses the principles of kirigami, Japanese paper-cutting, to design retractable belly scales that can help the robot grip the ground. Despite such advances, it will be quite some time before soft robots are used in commercial or industrial settings. Unlike hard robots, soft robots do not yet have underlying principles in their design. Soft animals may reveal principles to bridge this gap.

Robots are not just getting softer. They are getting smaller as well. The computer age was made possible by microfabrication techniques such as photo lithography, the use of light to etch designs in circuit boards. This technology is now being applied to fabricate small moving robot parts. The resulting micro-robots are finding application in medicine, such as in the camera pill, and in the military, such as in small flying surveillance robots. Whether these flying micro-robots rely upon either propellers or wings, they are subject to the same challenges that insects face. They must fly quickly, deal with incoming flows, and land on a variety of surfaces. A greater understanding of insect flight, and especially insect visual systems, may be needed before robotic insects take to the skies.

Some might consider there to be downsides to imbuing robots with the ability to move like animals. Indeed, robotic flapping birds are so lifelike that they are difficult to distinguish from far away. Engineers such as Mark Cutkosky at Stanford and Russ Tedrake at MIT are helping flying robots perch on branches, giving them even more ability to maneuver and land in outdoor environments. We don't typically give a second thought if animals around us are watching. If a robotic bird perches on our window, or a robotic fly lands on our desk, how can we ensure our privacy? These are important questions, and as robots become more animal-like, we will need associated regulations. The Federal Aviation Administration has already made similar measures for drones, and they might oversee flapping vehicles as well.

Along with its promising future, the study of animal movement also faces challenges. Biology departments are seeing a rise in scientists who study molecular, cellular, and systems biology, which in turn are displacing traditional biologists, especially those who study whole-animal motion. A biologist today is far more likely to spend a lifetime on a certain number of model organisms such as the mouse, fruit fly, or nematode. Consequently, most of the diversity in nature is no longer studied at the scope it once was. For example, taxonomists, who are trained in finding and identifying species are, ironically, going extinct themselves. Fewer active taxonomists leave fewer to train the next generation. To survive, taxonomists are beginning to train their next generation with the tools of genetics to better integrate themselves into modern biology departments. Without taxonomists, it will be increasingly difficult to recognize the effect of humans on diversity and numbers of animals in the wild.

Much of my correspondence with students concerns how to find careers or jobs in animal motion. My own position at Georgia Tech, a joint faculty appointment in mechanical engineering and biology, was not advertised, but was specially created for me. A number of my colleagues are in a similar situation. They were able to sell their interests in animal motion in the context of the broad aims of computer science, materials science, and physics.

Other places remain safe havens for those studying whole animals. In medical and veterinary schools, positions exist for experts on anatomy because dissection remains the primary method that anatomy is taught. Few locations have their facilities for observation of large animals. In universities without medical or veterinary schools, core biology courses focus less on anatomy, but more on genetics and molecular biology. In such places, an interest in animal motion would need to be linked to other fields of study.

Museums, aquariums, and zoos are also important employers of those interested in animal locomotion science. Joe Mendelson, who I've worked with for years, is director of research at the Atlanta Zoo, where he works across departments on research topics from biomechanics to conservation. Institutions like Joe's have long histories of doing research and hosting visiting scientists around the world to do research. Positions like Joe's also give him the luxury of working with live animals. Much of my own research would not be possible without partnership with the knowledgeable staff at the Atlanta Zoo.

* * *

The public has an important role to play in helping the field of animal motion. According to UNESCO, the number of scientists in the world is seven million: one for every 1000 citizens. Each citizen has the potential to be a citizen scientist, contributing data to a study on animal movement. Many have access to a cell phone with a first-rate video camera. Social media and video-sharing websites such as YouTube have a profound effect on animal movement studies by communicating sightings of animals across the world. Much of the data from my work on animal urination was found in online videos uploaded by citizen scientists. In a 2010 program called "Flight artists," Dutch mechanical engineer David Lentink loaned a number of high-speed cameras to Dutch citizens to film the flight of birds and insects. The Cornell Lab of Ornithology helps the bird-watching community through a highly successful cell phone app. It allows citizens to input bird sightings, and automatically estimates distributions of bird species across the United States. These are excellent examples of crowd-sourcing. There is always a need for more volunteers. Contact your local zoo, natural history museum, or professors at your local university to ask how you can help.

Aristotle thought there was a soul in the animal that made it alive. This idea, called vitalism, has since been discounted, in favor of determinism, in which animals are simply highly complex systems of working biological parts. In fact, one of the feats of modern biology is showing that all life is composed of cells, and those cells of molecules. If animals are simply very complex machines, then so are we, making everything we do potentially replicable one day using inanimate parts. We are still a long away from ever doing so. But reading this book is one step along your personal journey, toward understanding why life is so remarkable. I hope you agree that understanding can give rise to appreciation without detracting from beauty. There are many who believe that if we take apart something, we make it less beautiful. Those people I believe have not had the chance to be on the other side, in my shoes, and now yours, where trying to understand things gives a secret joy and even greater appreciation of the world around us. Although this marks the end of the book, I hope it marks your beginning as you see your world and the animals in it a little differently, with an eye for the details, a heart open to strangeness and wonder, and a commitment to always asking why.

Bibliography

INTRODUCTION

Dickerson, Andrew K., Zachary G. Mills, and David L. Hu. 2012. "Wet Mammals Shake at Tuned Frequencies to Dry." *Journal of the Royal Society Interface* 9 (77):3208–18.

Fish, Frank E. 2006. "The Myth and Reality of Gray's Paradox: Implication of Dolphin Drag Reduction for Technology." *Bioinspiration and Biomimetics* 1 (2):R17.

Gray, J. 1957. "How Fishes Swim." *Scientific American* 197 (August):48–54.

Thompson, D. W. 1942. *On Growth and Form.* Cambridge University Press.

CHAPTER 1

Barthlott, Wilhelm, and Christoph Neinhuis. 1997. "Purity of the Sacred Lotus, or Escape from Contamination in Biological Surfaces." *Planta* 202 (1): 1–8.

Bush, John W. M., and David L. Hu. 2006. "Walking on Water: Biolocomotion at the Interface." *Annual Review of Fluid Mechanics* 38:339–69.

Bush, John W. M., David L. Hu, and Manu Prakash. 2007. "The Integument of Water-Walking Arthropods: Form and Function." In *Advances in Insect Physiology*, edited by J. Casas and S. J. Simpson, 34:117–92. Academic Press.

Hu, David L., Brian Chan, and John W. M. Bush. 2003. "The Hydrodynamics of Water Strider Locomotion." *Nature* 424 (6949):663–66.

Koh, Je-Sung, Eunjin Yang, Gwang-Pil Jung, Sun-Pill Jung, Jae Hak Son, Sang-Im Lee, Piotr G. Jablonski, Robert J. Wood, Ho-Young Kim, and Kyu-Jin Cho. 2015. "Jumping on Water: Surface Tension–Dominated Jumping of Water Striders and Robotic Insects." *Science* 349 (6247):517.

Van Dyke, Milton. 1982. *An Album of Fluid Motion.* Parabolic Press.

Wang, Qianbin, Xi Yao, Huan Liu, David Quéré, and Lei Jiang. 2015. "Self-Removal of Condensed Water on the Legs of Water Striders." *Proceedings of the National Academy of Sciences of the United States of America* 112 (30):9247–52.

CHAPTER 2

Darwin, Charles, and Francis Darwin. 1898. *The Formation of Vegetable Mould through the Action of Worms*. D. Appleton.

Dorgan, Kelly M., Sanjay R. Arwade, and Peter A. Jumars. 2008. "Worms as Wedges: Effects of Sediment Mechanics on Burrowing Behavior." *Journal of Marine Research* 66 (2):219–54.

Gettelfinger, Brian, and E. L. Cussler. 2004. "Will Humans Swim Faster or Slower in Syrup?" *AIChE Journal* 50 (11): 2646–47.

Hu, David L., Jasmine Nirody, Terri Scott, and Michael J. Shelley. 2009. "The Mechanics of Slithering Locomotion." *Proceedings of the National Academy of Sciences* 106 (25):10081.

Jones, William J., Shannon B. Johnson, Greg W. Rouse, and Robert C. Vrijenhoek. 2008. "Marine Worms (Genus *Osedax*) Colonize Cow Bones." *Proceedings of the Royal Society B: Biological Sciences* 275 (1633):387–91.

Li, Chen, Paul B. Umbanhowar, Haldun Komsuoglu, Daniel E. Koditschek, and Daniel I. Goldman. 2009. "Sensitive Dependence of the Motion of a Legged Robot on Granular Media." *Proceedings of the National Academy of Sciences* 106 (9):3029.

Maladen, Ryan D., Yang Ding, Chen Li, and Daniel I. Goldman. 2009. "Undulatory Swimming in Sand: Subsurface Locomotion of the Sadfish Lizard." *Science* 325 (5938):314–18.

Quillin, K. J. 2000. "Ontogenetic Scaling of Burrowing Forces in the Earthworm *Lumbricus terrestris*." *Journal of Experimental Biology* 203 (18):2757–70.

CHAPTER 3

Dabiri, John O., S. P. Colin, K. Katija, and John H. Costello. 2010. "A Wake-based Correlate of Swimming Performance and Foraging Behavior in Seven Co-occurring Jellyfish Species." *Journal of Experimental Biology* 213 (8):1217–25.

Dabiri, John O., and Morteza Gharib. 2005. "The Role of Optimal Vortex Formation in Biological Fluid Transport." *Proceedings of the Royal Society B: Biological Sciences* 272 (1572):1557.

Garcia, Guilherme J. M., and Jafferson Kamphorst Leal da Silva. 2006. "Interspecific Allometry of Bone Dimensions: A Review of the Theoretical Models." *Physics of Life Reviews* 3 (3):188–209.

Gharib, Morteza, Edmond Rambod, and Karim Shariff. 1998. "A Universal Time Scale for Vortex Ring Formation." *Journal of Fluid Mechanics* 360:121–40.

Haldane, John B. S. 1926. "On Being the Right Size." *Harper's Magazine* 152 (March):424–27.

Holden, Daniel, John J. Socha, Nicholas D. Cardwell, and Pavlos P. Vlachos. 2014. "Aerodynamics of the Flying Snake *Chrysopelea paradisi*: How a Bluff Body Cross-sectional Shape Contributes to Gliding Performance." *Journal of Experimental Biology* 217 (3):382–94.

Ruiz, Lydia A., Robert W. Whittlesey, and John O. Dabiri. 2011. "Vortex-Enhanced Propulsion." *Journal of Fluid Mechanics* 668:5–32.

Schmidt-Nielsen, Knut. 1984. *Scaling, Why Is Animal Size So Important?* Cambridge University Press.

Vogel, Steven. 2003. *Comparative Biomechanics: Life's Physical World.* Princeton University Press.

Yang, Patricia J., Jonathan Pham, Jerome Choo, and David L. Hu. 2014. "Duration of Urination Does Not Change with Body Size." *Proceedings of the National Academy of Sciences* 111 (33):11932–37.

CHAPTER 4

Amador, Guillermo J., and David L. Hu. 2015. "Cleanliness Is Next to Godliness: Mechanisms for Staying Clean." *The Journal of Experimental Biology* 218 (20):3164.

Amador, Guillermo J., Wenbin Mao, Peter DeMercurio, Carmen Montero, Joel Clewis, Alexander Alexeev, and David L. Hu. 2015. "Eyelashes Divert Airflow to Protect the Eye." *Journal of the Royal Society Interface* 12 (105):20141294.

Levy, Y., N. Segal, D. Ben-Amitai, and Y. L. Danon. 2004. "Eyelash Length in Children and Adolescents with Allergic Diseases." *Pediatric Dermatology* 21 (5): 534–37.

Oeffner, Johannes, and George V. Lauder. 2012. "The Hydrodynamic Function of Shark Skin and Two Biomimetic Applications." *Journal of Experimental Biology* 215 (5):785–95.

Reif, W. 1985. *Squamation and Ecology of Sharks.* Senckenbergische Naturforschende Gesellschaft.

Wen, Li, James C. Weaver, and George V. Lauder. 2014. "Biomimetic Shark Skin: Design, Fabrication and Hydrodynamic Function." *Journal of Experimental Biology* 217 (10):1656–66.

CHAPTER 5

Collins, Steve, Andy Ruina, Russ Tedrake, and Martijn Wisse. 2005. "Efficient Bipedal Robots Based on Passive-Dynamic Walkers." *Science* 307 (5712):1082.

Collins, Steven H., M. Bruce Wiggin, and Gregory S. Sawicki. 2015. "Reducing the Energy Cost of Human Walking Using an Unpowered Exoskeleton." *Nature* 522 (7555):212–15.

Dickinson, Michael H., Claire T. Farley, Robert J. Full, M.A.R. Koehl, Rodger Kram, and Steven Lehman. 2000. "How Animals Move: An Integrative View." *Science* 288 (5463):100.

Liao, James C., David N. Beal, George V. Lauder, and Michael S. Triantafyllou. 2003. "Fish Exploiting Vortices Decrease Muscle Activity." *Science* 302 (5650):1566–69.

Matthis, Jonathan Samir, and Brett R. Fajen. 2013. "Humans Exploit the Biomechanics of Bipedal Gait During Visually Guided Walking over Complex Terrain." *Proceedings of the Royal Society B: Biological Sciences* 280 (1762):20130700.

Roberts, Thomas J., Richard L. Marsh, Peter G. Weyand, and C. Richard Taylor. 1997. "Muscular Force in Running Turkeys: The Economy of Minimizing Work." *Science* 275 (5303):1113.

CHAPTER 6

Dickerson, Andrew K., Peter G. Shankles, Nihar M. Madhavan, and David L. Hu. 2012. "Mosquitoes Survive Raindrop Collisions by Virtue of Their Low Mass." *Proceedings of the National Academy of Sciences* 109 (25):9822–27.

Foster, D. J. and R. V. Cartar. 2011. "What Causes Wing Wear in Foraging Bumble Bees? *Journal of Experimental Biology* 214(11): 1896–1901.

Jayaram, Kaushik, and Robert J. Full. 2016. "Cockroaches Traverse Crevices, Crawl Rapidly in Confined Spaces, and Inspire a Soft, Legged Robot." *Proceedings of the National Academy of Sciences* 113 (8):E950–57.

Mountcastle, Andrew M., and Stacey A. Combes. 2013. "Wing Flexibility Enhances Load-Lifting Capacity in Bumblebees." *Proceedings of the Royal Society B: Biological Sciences* 280 (1759).

Mountcastle, Andrew M., and Stacey A. Combes. 2014. "Biomechanical Strategies for Mitigating Collision Damage in Insect Wings: Structural Design versus Embedded Elastic Materials." *Journal of Experimental Biology* 217 (7):1108.

CHAPTER 7

Chang, Song, and Z. Jane Wang. 2014. "Predicting Fruit Fly's Sensing Rate with Insect Flight Simulations." *Proceedings of the National Academy of Sciences* 111 (31):11246–51.

Cowan, Noah J., Jusuk Lee, and R. J. Full. 2006. "Task-level Control of Rapid Wall Following in the American Cockroach." *Journal of Experimental Biology* 209 (9):1617–29.

Grillner, Sten. 1996. "Neural Networks for Vertebrate Locomotion." *Scientific American* 274 (1):64–69.

Ijspeert, Auke Jan, Alessandro Crespi, Dimitri Ryczko, and Jean-Marie Cabelguen. 2007. "From Swimming to Walking with a Salamander Robot Driven by a Spinal Cord Model." *Science* 315 (5817):1416–20.

Lee, Jusuk, Simon N. Sponberg, Owen Y. Loh, Andrew G. Lamperski, Robert J. Full, and Noah J. Cowan. 2008. "Templates and Anchors for Antenna-based Wall Following in Cockroaches and Robots." *IEEE Transactions on Robotics* 24 (1):130–43.

Muijres, Florian T., Michael J. Elzinga, Johan M. Melis, and Michael H. Dickinson. 2014. "Flies Evade Looming Targets by Executing Rapid Visually Directed Banked Turns." *Science* 344 (6180):172–77.

Ristroph, Leif, Attila J. Bergou, Gunnar Ristroph, Katherine Coumes, Gordon J. Berman, John Guckenheimer, Z. Jane Wang, and Itai Cohen. 2010. "Discovering the Flight Autostabilizer of Fruit Flies by Inducing Aerial Stumbles." *Proceedings of the National Academy of Sciences* 107 (11):4820.

Wang, Z. Jane. 2000. "Vortex Shedding and Frequency Selection in Flapping Flight." *Journal of Fluid Mechanics* 410:323–41.

CHAPTER 8

Garnier, Simon, Jacques Gautrais, and Guy Theraulaz. 2007. "The Biological Principles of Swarm Intelligence." *Swarm Intelligence* 1 (1):3–31.

Mlot, Nathan J., Craig A. Tovey, and David L. Hu. 2011. "Fire Ants Self-assemble into Waterproof Rafts to Survive Floods." *Proceedings of the National Academy of Sciences* 108 (19):7669–73.

Parrish, Julia K., and Leah Edelstein-Keshet. 1999. "Complexity, Pattern, and Evolutionary Trade-Offs in Animal Aggregation." *Science* 284 (5411):99.

Reid, Chris R., Matthew J. Lutz, Scott Powell, Albert B. Kao, Iain D. Couzin, and Simon Garnier. 2015. "Army Ants Dynamically Adjust Living Bridges in Response to a Cost-Benefit Trade-off." *Proceedings of the National Academy of Sciences* 112 (49):15113–18.

Rubenstein, Michael, Alejandro Cornejo, and Radhika Nagpal. 2014. "Programmable Self-Assembly in a Thousand-Robot Swarm." *Science* 345 (6198):795–99.

Sumpter, D.J.T. 2006. "The Principles of Collective Animal Behaviour." *Philosophical Transactions of the Royal Society B: Biological Sciences* 361 (1465):5.

Yim, M., W.-m. Shen, B. Salemi, D. Rus, M. Moll, H. Lipson, E. Klavins, and G. S. Chirikjian. 2007. "Modular Self-Reconfigurable Robot Systems [Grand Challenges of Robotics]." *IEEE Robotics and Automation Magazine* 14 (1):43–52.

CONCLUSION

Amador, Guillermo J., and David L. Hu. 2015. "Cleanliness Is Next to Godliness: Mechanisms for Staying Clean." *Journal of Experimental Biology* 218 (20):3164–74.

Brainerd, Elizabeth L., David B. Baier, Stephen M. Gatesy, Tyson L. Hedrick, Keith A. Metzger, Susannah L. Gilbert, and Joseph J. Crisco. 2010. "X-Ray Reconstruction of Moving Morphology (XROMM): Precision, Accuracy and Applications in Comparative Biomechanics Research." *Journal of Experimental Zoology Part A: Ecological Genetics and Physiology* 313A (5):262–79.

Chittka, Lars, and Jeremy Niven. 2009. "Are Bigger Brains Better?" *Current Biology* 19 (21):R995–R1008.

Hu, David L. 2016. Confessions of a Wasteful Scientist. *Scientific American.* (Guest blog.)

Khurram, Abeer, Shani E. Ross, Zachariah J. Sperry, Aileen Ouyang, Christopher Stephan, Ahmad A. Jiman, and Tim M. Bruns. 2017. "Chronic Monitoring of Lower Urinary Tract Activity via a Sacral Dorsal Root Ganglia Interface." *Journal of Neural Engineering* 14 (3): 036027.

Kim, Sangbae, Cecilia Laschi, and Barry Trimmer. 2013. "Soft Robotics: A Bio-inspired Evolution in Robotics." *Trends in Biotechnology* 31 (5):287–94.

Ma, Kevin Y., Pakpong Chirarattananon, Sawyer B. Fuller, and Robert J. Wood. 2013. "Controlled Flight of a Biologically Inspired, Insect-Scale Robot." *Science* 340 (6132):603.

Park, Sung-Jin, Mattia Gazzola, Kyung Soo Park, Shirley Park, Valentina Di Santo, Erin L. Blevins, Johan U. Lind, et al. 2016. "Phototactic Guidance of a Tissue-Engineered Soft-Robotic Ray." *Science* 353 (6295):158.

Rus, Daniela, and Michael T. Tolley. 2015. "Design, Fabrication and Control of Soft Robots." *Nature* 521 (May):467.

Schwenk, Kurt, Dianna K. Padilla, George S. Bakken, and Robert J. Full. 2009. "Grand Challenges in Organismal Biology." *Integrative and Comparative Biology* 49 (1):7–14.

Versteegden, Luuk R., Kenny A. van Kampen, Heinz P. Janke, Dorien M. Tiemessen, Henk R. Hoogenkamp, Theo G. Hafmans, Edwin A. Roozen, et al. 2017. "Tubular Collagen Scaffolds with Radial Elasticity for Hollow Organ Regeneration." *Acta Biomaterialia* 52:1–8.

Index